Finite Element Analysis of Prestressed Concrete Structures Using Post-Tensioning Steel

Finite Element Analysis of Prestressed Concrete Structures Using Post-Tensioning Steel

By

Yu Huang and Thomas Kang

Cambridge Scholars Publishing

Finite Element Analysis of Prestressed Concrete Structures Using Post-Tensioning Steel

By Yu Huang and Thomas Kang

This book first published 2020

Cambridge Scholars Publishing

Lady Stephenson Library, Newcastle upon Tyne, NE6 2PA, UK

British Library Cataloguing in Publication Data
A catalogue record for this book is available from the British Library

ISBN (10): 1-5275-4285-8
ISBN (13): 978-1-5275-4285-3

TABLE OF CONTENTS

LIST OF TABLES

LIST OF FIGURES

ACKNOWLEDGEMENTS

This book is based on the PhD dissertation of the first author in 2012, with significant updates that have been made until 2020. The research herein would not have been possible without the practical supports of numerous people. Thus the first author's sincere gratitude goes to his advisor, his Ph.D. committee members, his parents and all his friends for their love, support and patience over the last few years.

Firstly, the first author owns his deepest gratitude to his brilliant and talented advisor Dr. Thomas Kang, the second author of this book, for encouragement, supervision, and advisement during the Ph.D. career. This research is next to impossible without Dr. Kang's expert guidance and support. Not only was he readily available for me and other students as he always responded to the first author's writings and research materials promptly and actively.

The first author's thanks also go to Dr. Chris Ramseyer, one of the first author's PhD committee, who has been so supportive during the studies at the University of Oklahoma. It is a pleasure to acknowledge dear other Ph.D. committee members, Dr. James Baldwin, Dr. S. Lakshmivarahan, and Dr. Lisa Holliday, for their expert insights, encouragement and guidance throughout the whole research. Many people on the faculty and staff of the College of Engineering and other outside departments at the University of Oklahoma assisted and encouraged the first author in various ways during the course of studies. The first author is particularly grateful to Dr. Jinsong Pei and Dr. Kyran Mish for all that they have taught the first author and support of his Ph.D. study.

Also, the first author would like to credit all of his fellow student colleagues in the School of Civil Engineering and Environmental Science at the University of Oklahoma. The first author's graduate studies would not have been so colorful and exciting without the social and academic challenges and diversities provided by them. I especially want to thank Dr. Woosuk Kim, an Assistant Professor at Kumoh National Institute of Technology, who always offered his insights and helps for the first author's study.

In addition, the first author wishes to express his warmest gratitude to his family in China for their unconditional support and continuous encouragement during the Ph.D. career.

Finally, the first author wishes to thank all whose support, guidance, and encouragement from the preliminary to the concluding level of the research have enabled him to develop systematic skills and a thorough understanding of the subject. The authors remain responsible for the content of the following including all unwittingly remained errors.

It is also noted that the unit primarily used in this book is the customary US unit, with conversion to the SI unit available in some of the figures.

ABSTRACT

This research discusses and investigates possible approaches for modeling post-tensioned (PT) prestressed concrete structures via the finite element method (FEM). The challenge of modeling PT prestressed concrete structures lies in the treatment of the interface between the concrete and prestressing tendons. The generic modeling techniques that are discussed were based on general purpose finite element packages. Two strategies for modeling the interface are presented in detail. For the first method, a series of linear spring elements was introduced to approximate the sliding behavior of prestressing tendons at the interface. For the second method, the interface was modeled directly through contact formulation. Additionally, the corresponding material constitutive relations, element preference and solution algorithm are discussed in depth. The generic modeling schemes were validated against experiments and proved to be robust and reliable for modeling PT structures. However, slightly overestimated tendon stresses were observed at the ultimate state of structures in many numerical simulations. A preliminary scientific analysis yields the conclusion that the phenomenon is likely caused by the frictionless assumption which neglects the frictional-induced prestress loss in the modeling.

Besides the issue of overestimated prestressing tendon stress, it is difficult to apply the generic modeling schemes to simulate large-scale PT structures such as a PT frame system. The involvement of the solid element combined with the explicit dynamic algorithm becomes a large barrier to modeling large-scale structures due to the computational anxiety. The later part of the research zeros in on the development of an innovative nonlinear finite element formulation which incorporates contact techniques and engineering elements to considerably reduce the need of computational power. A nonlinear prototype program was developed to model PT prestressed concrete frames in two-dimensional space accordingly. The stress solution of prestressing tendons was also improved by considering frictional effects in the formulation. The proposed formulation was validated against analytic solutions and experimental data via several numerical studies. The prototype program was also demonstrated to be versatile and robust for analyzing PT prestressed concrete frames. Although the currently implemented material constitutive relations and beam element in the prototype program limit its applications, advanced material models and beam elements could be implemented into the current formulation with trivial works in the future.

In addition to the study of modeling techniques, three practical engineering problems were investigated through the proposed FEM. The investigated problems include: (1) punching shear failure of two-way PT prestressed concrete slabs; (2) prestress increment in the prestressing tendon at the service stage of structural members; and (3) influence of the PT systems (i.e., bonded vs. unbonded) on structural performance of typical PT members. A series of numerical simulations based on general purpose finite element packages were conducted according to the documented experiments. The extensive analyses of the numerical and experimental data lead to the following conclusions: (1) the eccentric shear stress model proposed in ACI 318 predicts reasonable moment-shear interaction mechanisms for PT interior, edge, and, corner slab-column connections; (2) in the case of PT edge and corner connections, the punching shear provisions in ACI 318 are overly conservative in some cases, more research is suggested to quantify the prestressing effect on punching shear capacity and to relieve some of the provisions; and (3) the bonding condition of prestressing tendons has no effect on flexural strength of PT one-way slabs and beams, or the moment-shear interaction in two-way PT interior slab-column connections.

CHAPTER ONE

INTRODUCTION

Prestressed concrete is a structural concept that was first introduced for overcoming concrete's natural weakness in tension in the early 1940s. Prestressed concrete is essential in many applications today in order to fully utilize concrete compressive strength, and through proper design, to control cracking and deflection. Due to those benefits, the prestressed concrete industry has been experiencing breathtaking developments and a construction boom ever since the first prestressed concrete project, Walnut Lane Memorial Bridge, was built in Philadelphia, Pennsylvania in 1951. The efforts toward development made by many engineers over the past sixty years have made prestressed concrete an essential and powerful technique in the modern construction industry. Generally, prestressed concrete can be applied in one of three ways: as pre-tensioned prestressed concrete; as unbonded post-tensioned (PT) prestressed concrete and as bonded PT prestressed concrete. The first two have become the most popular types of prestressed concrete systems in North America.

Although design methods have been developed over the decades, an understanding of the ultimate mechanism in the prestressed concrete system is still greatly needed in many aspects. Such aspects include the intricate problems of punching shear failure of a prestressed two-way slab system, stress increase in the prestressing strand under service loads, and discrepancies of structural behaviors between bonded and unbonded post-tensioned systems. However, to perform extensive experimental tests on each subject is extremely expensive and time-consuming. The finite element method, on the other hand, was introduced into structural analysis in the late 1960s. The efforts and developments made by many pioneering researchers over the past five decades have enabled the finite element method to become a versatile and powerful approach in structural analysis. The principal goals of this study are establishing finite element models of prestressed concrete systems subject to particular problems and investigating those problems in accordance with current building codes.

Chapter 2 summarizes previous experimental tests regarding particular engineering problems. The problems considered include punching shear failure of two-way post-tensioned slabs, stress increase in prestressing tendons at the failure of the structural members and an investigation about the influence on the behaviors of prestressed concrete systems by different bonding (i.e., bonded PT vs. unbonded PT). Related literatures of finite element modeling with respect to prestressed concrete systems are provided.

Chapter 3 briefly presents the hypothesis and objectives of the book.

Chapter 4 provides descriptions of modeling prestressed concrete systems via existing general-purpose finite element packages. The modeling of PT tendon systems presents a great challenge in terms of finite element discretization due to the boundary nonlinearity. Two possible approaches to modeling PT tendon systems are discussed. The first and traditional approach is a so called ‘spring system method’ which literally utilizes a large number of rigid springs to approximate the mechanical behavior of an unbonded PT system. The second approach formulates the PT tendon systems as contact problems. The nonlinear boundaries are directly handled by the second approach which yields a better flexibility and accuracy of modeling a variety of PT structures. The spring method is found in many literatures whereas the contact formulation is rarely applied to finite element analysis in the context of PT structures. The reason behind this is that the generic contact problems are difficult to model and still remains a very active research area today. In addition, directly formulating nonlinear boundary problems via contact approach usually requires much more computational power than the traditional spring method.

Followed by the general modeling approaches discussed in Chapter 4, a large number of documented tests are numerically simulated to validate the proposed modeling approaches as well as to conduct further researches in Chapter 5. The numerically modeled specimens (19 slabs and 3 beams) are grouped to two parts where the first part of the simulations is used to analyze the punching shear failure of a prestressed two-way slab system. The second part is used to study discrepancies of structural behaviors between bonded and unbonded PT systems. The numerical results are well agreed with experimental data and are used to investigate the aforementioned practical engineering problems.

The drawback of using a general-purpose finite element package is that one has less modeling flexibility. For example, prestress loss is hard to model at several prestressing stages. Therefore, a two-dimensional nonlinear finite element formulation incorporating contact techniques is proposed in Chapter 6. The formulation can analyze inelastic behavior of PT beams. Several elements are developed in order to assemble the complete PT beam system. The proposed contact element deals with the interaction between concrete and prestressing tendons, the nonlinear beam element simulates conventional reinforced concrete (RC) beams; the nonlinear truss element assembles the prestressing tendon; the embedding element embeds prestressing tendons into concrete beams, and a special anchorage element is also proposed to accommodate the ability of simulating post-tensioning procedures. With all these elements assembled together, a complete PT beam system can be successfully modeled along with appropriate material constitutive models. The solution strategy employs a modified Newton-Raphson approach with a line search technique that improves the robustness of the method. Accordingly, all tangent stiffness matrices of aforementioned elements are analytically derived in accordance with the proposed nonlinear solver. Both frictionless and frictional contacts are formulated in simulating unbonded and bonded PT beams. Furthermore, in the proposed finite element framework it is easy to implement other types of elements such as elements with higher order beam theory (e.g., a Timoshenko beam element) to address different kinds of practical engineering problems. Enough accuracy can be obtained with a relatively coarse mesh under the current formulation to compare to the modeling approach with solid elements. This leads to a very cost-efficient solution to studying engineering problems and aid practical designs. All the proposed formulations are programmed and implemented by MATLAB language (MATLAB, 2010b). Several numerical studies are carried out to inspect the performance of the proposed model.

In Chapter 7, all materials are summarized together. Conclusions are given regarding finite element modeling techniques of PT structures as well as the practical engineering problems analyzed and evaluated by the proposed modeling scheme.

CHAPTER TWO

INVESTIGATION OF EXISTING STUDIES

2.1 Punching Shear Failure of PT Slab-Column Connections

The following review covers previous experiments conducted regarding PT slab-column connections or PT flat plates under quasi-static gravity and/or lateral loadings. Some of them were selected as the experimental basis of the following numerical analyses in Chapter 5.

Scordelis et al. (1959) tested a two-way unbonded PT flat plate system with four panels supported at nine points (eight points along edges and one point at the center). The slab was post-tensioned with twelve cables in each direction, uniformly distributed with a draped parabolic profile to balance desired moments. Each cable consisted of a single 1/4 in. high strength steel wire greased and placed in a plastic sheathing. Four tests were performed, one each under uniform prestress; under unequal prestress (1.8:1 for column to middle strip), skip loading (live load was only presented on one panel) and uniform loading to failure (live load was presented on four panels). A punching shear failure occurred at the center support after extensive flexural cracking. Theoretical calculations were also performed to predict moment and deflection within the elastic range by the beam theory and elastic plate method. The design method was evaluated by the experiment.

Brotchie and Beresford (1967) conducted an experimental test of an unbonded PT flat plate system. The overall slab dimensions were 26 ft by 42 ft in footprint and 3 in. in thickness, with supporting columns spaced on a 12 ft by 9 ft grid. Prestressing tendons were single 0.276 in. wires unbonded and draped to follow overall panel moment profiles and tensioned to balance these moments at the sustained loading. A 30-month long test was performed under sustained loading followed by a short term uniform loading test until flexural failure occurred. The slab system failed by a process of folding and yield rotation initialization where actual shear capacity was unknown.

An experimental study followed Scordelis' study (1959) and was performed by Odello and Mehta (1967). The test specimen was identical to the specimen in Scordelis' study except that five drop panels were added at the edge and center supports, respectively. The slab was loaded and unloaded at several levels and monotonically loaded to failure. Flexural failure was observed first followed by an ultimate failure of combined flexure and shear failure with a slightly higher load. The beam theory was investigated with this experiment and yielded a satisfactory elastic and ultimate analysis. Cracking and ultimate load carrying capacities of the slab were increased significantly with the presence of drop panels. These three experiments were related to large scale unbonded PT flat plate systems. However, conclusions on the punching shear capacity of slab-column connections under such prestressed concrete systems have not been reached.

Other researchers (Gamble, 1964; Burns and Hemakom, 1977; Burns and Hemakom, 1985; Kosut et al., 1985) conducted experiments on unbonded PT flat plate systems with attempts to investigate the punching shear strength. A two bay by three bay continuous unbonded PT flat plate was tested by Gamble (1964). The plate system consisted of two spans of 12 ft in the transverse direction and three 9 ft spans in the longitudinal direction along with twelve square column supports underneath the slab. The lightweight concrete slab was post-tensioned in both directions using 0.276 in. diameter high strength wire with a straight profile. The wires were spaced at 4 in. in the direction of the 12 ft span and 6 in. in the direction of the 9 ft span. Additional non-prestressed mild steel was also provided. Long term and short term tests were performed. The specimen was loaded uniformly until all panels failed. A very brittle shear failure occurred at one of the interior slab-column connections before flexural failure. The moment distribution was investigated and was in a reasonable agreement with the elastic moment before cracking. The flexural strength was predicted by using the yield line theory but was not able to be verified by the experiment. Punching shear strength was evaluated based on several available equations for the interior slab-column connection, whereas no investigation was made to the corner connections.

Burns and Hemakom (1977) tested a one-third scale unbonded PT flat plate with nine panels. The slab was three bay by three bay with 10 ft spans in each direction and was post-tensioned with 68 1/4 in. diameter seven-wire strands in each direction with a draped profile. The tendons were distributed 70% in the column strip and 30% in the middle strip. In addition, non-prestressed mild steel was provided at the column regions. The slab was loaded by a whiffletree system producing sixteen load points on each main panel and four load points on each span of overhang. A total of 15 tests were performed. The first test was about the instrument check. Nine of them were related to flexural tests and five of them were subjected to punching shear failure tests. The experimental observations indicated that the shear capacity was relatively consistent from column to column. The moment distribution observed from the experiments was compared with the elastic plate theory and equivalent frame method, indicating the design method performed well. In addition, the flexural capacity was accurately predicted by the yield line theory. Another half scale unbonded PT flat plate with nine panels was tested by Burns and Hemakom (1985). The slab had three spans of 10 ft in each direction and was post-tensioned with 23 1/4 in. diameter tendons in the north-south direction and 24 1/4 in. diameter tendons in the west-east direction. The tendons were uniformly distributed in the north-south direction and banded in the column strip in the west-east direction. All tendons had a draped profile. Additional non-prestressed mild steel was provided at the column regions. A similar loading procedure was employed in a total of 12 tests. The first test was performed to check instruments. Eight of the tests were flexural tests and three of the tests were punching shear tests. In all failure load tests, the slab failed by flexure followed by a punching shear failure with the same load. Burns and Hemakom (1985) concluded a large deflection and curvature at the negative moment yield line might have triggered the shear failure after the flexural failure.

Kosut et al. (1985) conducted a test of half scale unbonded PT flat plate with four panels. The test slab was 20 ft square with two spans of 10 ft in each direction and nominally 2.75 in. in thickness. The 0.25 in. diameter, 244 ksi strength wires were uniformly distributed in one direction and banded in the column strip in the other direction. All tendons had a draped profile. Auxiliary non-prestressed bonded reinforcement was provided at the connection. Several connections were further reinforced by shear stirrups. A similar loading system used by Burns and Hemakom (1977; 1985) was employed in this test. A total of 13 tests were performed of which eight were related to flexural tests and four were tested to assess the shear strength of individual slab-column connections. Test results showed use of vertical reinforcement at the exterior slab-column connection did not increase the shear strength.

Martinez-Cruzado (1993) investigated two 3/7 scale isolated unbonded PT edge slab-column connections and two corner connections. The slabs with the corner connection had an overall length of 7 ft 1-1/2 in. in each direction and 3-5/8 in. in thickness. The corner connection was in the south-west corner, whereas other corners of the slab were supported by pin connections simulating an inflection boundary in the prototype structure. The slab with the edge connection had an overall length of 13 ft 3 in. in the lateral loading direction and 6 ft 11-11/32 in. in the transverse direction, and a slab thickness of 3-5/8 in. Similar boundary conditions were used for the edge connection specimens. Five 3/8 in. diameter prestressing strands were banded in the north-south direction concentrating in the column strip, while another two and three were banded and uniformly distributed inside and outside the column strip along the west-east direction, respectively, in the slabs. A similar arrangement of tendon layouts was used for the slab with the edge connection except the banded tendons were grouped in the west-east direction. All prestressing strands were drape shaped. Supplemental bonded reinforcing bars were provided at the negative moment region around the column. Additional dead loads were applied to the slab before testing to achieve the desired gravity load in the column at the initialization of the test. A cloverleaf displacement loading pattern was applied to the column top with several cycles of different drift ratios for the purpose of simulating multi-directional seismic loading. One of the conclusions reached in the study was that the presence of high compressive stress in the slab-column connection region increases the shear strength of the connection. The increase of tensile stress was very small in the strand under constant gravity and increased lateral load.

Gardner and Kallage (1998) tested a two bay by two bay unbonded PT flat plate. The slab had two spans of 8 ft 11-7/8 in. in each direction and 3.54 in. in thickness. Three of the edge columns were circular and the others were square shaped. The 0.51 in. diameter, greased, seven-wire strands were uniformly distributed in one direction and banded in the column strip in the other direction. All tendons had a draped profile. No other supplementary bonded reinforcement was provided. The slab was uniformly loaded to failure in increments with forty-point loads. The punching shear failure occurred in a circular edge slab-column connection through a uniformly distributed load increase. After the initial failure, additional supports were placed at the failed edge connection to further load until another punching shear failure occurred at a rectangular interior slab-column connection. By shoring the interior connection, load was applied again until a rectangular corner slab-column connection failed by punching shear. Gardner and Kallage (1998) proposed a method to predict the punching shear strength and concluded that the presence of precompression in a slab might affect the shear strength of edge and corner slab-column connections.

Foutch et al. (1990) tested four isolated PT slabs with edge connections. Four slabs had the same dimensions of 80 in. in one direction, 60 in. in the other direction and 4 in. in thickness. The slabs were only supported by the edge connections located at the center of the exterior edge. Slab1 and Slab2 had banded tendons perpendicular to the exterior edge while tendons were uniformly distributed in the other direction. Tendons were banded parallel to the exterior edge in Specimens Slab3 and Slab4. All tendons were Grade 270, 3/8 in. diameter, seven wire strands. In addition, non-prestressed mild steel was used as the crack control reinforcement in the vicinity of the column and was also placed as the top and bottom edge reinforcement around the perimeter of the slab to prevent the splitting crack. The loading pattern of Slab1 and Slab2 were different from Slab3 and Slab4. The loading positions varied by the moment-to-shear ratio. All slabs were monotonically loaded to a failure. It was supported by the test results that the shear strength of a slab-column edge connection benefited from the precompression in the concrete.

Warnitchai et al. (2004) tested a 3/5 scale bonded PT slab with an interior slab-column connection. The square slab had a length of 18 ft 8-2/5 in. along each direction and 4.72 in. in thickness. The column was right located at the center of the slab with a section dimension of 9.84 in. x 19.69 in. and extended 35.43 in. above and below the slab mid-plane. The slab was supported by pin connections at two sides in the direction of lateral loading. The clear span between the pin connections of two sides was 16 ft 4-4/5 in. The slab was post-tensioned with eight bonded straight, Grade 270, 1/2 in. diameter, seven-wire strands. The tendons were banded in the lateral loading direction and uniformly distributed in the other direction. Additional non-prestressed reinforcement was provided at the slab top and bottom. The top slab bars were concentrated only at the connection region. The bottom steel mat was provided throughout the whole slab. Additional gravity loads were applied to the slab by sand bags in order to obtain the desired gravity-to-shear ratio. The lateral load was applied on the top of the column with several cycles of different drift ratios until the punching shear failure occurred. The brittle failure occurred at the drift ratio of 2%.

Most of the experimental programs reviewed above, however, provided insufficient data to assess the punching shear strength in the PT slab-column connections. The experiments carried out by Martinez-Cruzado (1993), Foutch et al. (1990), and Warnitchai et al. (2004) were isolated systems which had enough data from tests. Therefore, the FEM models were constructed for those specimens. The descriptions of the modeling are addressed in Chapter 4.

2.2 Comparative Study of Prestressing Tendon Bonding

The following reviews previous investigation of bonding influence of prestressing tendons. They were selected as the experimental basis of the following numerical analyses in Chapter 5.

Comparative studies regarding the prestressing tendon bonding influence on the structural behavior have been rarely reported in the past. Mattock et al. (1971) conducted an experimental study of seven simple supported single-span beams and three continuous two-span beams. All beams were categorized into three groups for the test as three T-beams (CB1, CU1 and CU2), continuous over two spans of 28 ft each; three simply supported T-beams (TB1, TU1 and TU2) of 28 ft each; and three simply supported rectangular beams (RB1, RU1 and RU2) of 28 ft span each. An additional unbonded PT T-beam was tested which was identical to the TU1 and TU2, except that a single 3/8 in. diameter non-prestressed seven wire strand was provided as additional bonded reinforcement. Each of the beams was post-tensioned by two, Grade 270, 1/2 in. diameter seven wire strands. In the first beam of each group, the tendons were bonded by grouting after the post-tensioning. The tendons were left unbonded in the other two beams of each group. The tendons were draped parabolically in all simply supported beams with an effective depth of 10 in. at the mid-span and zero eccentricity at the ends of the beams. In all continuous beams, the tendons were draped with an effective depth of 10 in. at both the mid-spans and zero eccentricity at the ends of the beams. In addition, different non-prestressed bonded mild steel was provided for each beam in accordance with ACI 318-63 (1963). Dead weight of all test specimens was increased by 100 percent since all beams were half-scaled. Each span was subjected to four equal point loads which were applied at points 1.5 ft and 5.5 ft away from the mid-span on each side. The point loads were increased monotonically until the failure of test beams occurred. For the simply supported beams, it has been realized that the ultimate strength of the unbonded beams might have been up to 30% greater than that of the bonded beams even though they were with the same design strength according to ACI 318-63. The test results from both simply supported and continuous beams also revealed that the crack spacings of the unbonded beams were equivalent to or larger better than those of the bonded ones which implied good serviceability characteristics. Mattock et al. (1971) also concluded that the unbonded beams with non-prestressed bonded reinforcement behaved more like a flexural member than a hinged tied arch.

Cooke et al. (1981) investigated twelve simply supported PT one-way slabs. Nine of them were prestressed with the unbonded tendons and the rest were prestressed with the bonded tendons. Of the nine unbonded slabs, three (Slab1, Slab2 and Slab3) had a length of 15 ft 9 in. with spans of 15 ft 1-1/8 in.; three (Slab4, Slab5 and Slab6) had a length of 11 ft 9-3/4 in. with spans of 11 ft 1-3/4 in.; and three (Slab7, Slab8 and Slab9) had a length of 7 ft 10-1/2 in. with spans of 7 ft 2-5/8 in. The three bonded slabs (Slab B4, Slab B5 and Slab B6) were identical to Slab4, Slab5 and Slab6, respectively, except the tendons were bonded to the concrete. The slabs in each unbonded group had a width of 1 ft 1-7/8 in., 2 ft 3-3/4 in. and 3 ft 10-1/2 in. All slabs were 7-1/16 in. in the thickness. The first two slabs in each group were prestressed with three straight tendons of 1/2 in. diameter. The last slab in each group was prestressed with three straight tendons of 5/16 in. diameter. All tendons were placed at an effective depth of 4-3/4 in. No additional bonded reinforcement was provided to any of the slabs. All slabs were subjected to a constant bending moment by two line loads which were applied at 4 ft 11-1/16 in., 3 ft 7-5/16 in. and 2 ft 3-9/16 in., respectively, from the supports in each group. All slabs were statically loaded in increments until the failures occurred.

All equations predicting tendon ultimate stress were evaluated to have under-predicted the ultimate tendon stress except the equation from ACI 318-77 (1977). This equation yielded a slightly higher value compared to measured results only for a low prestressing steel index, q_e ($q_e = \rho_p f_{pe}/f'_c$). All the equations examined for flexural strength produced conservative values except those equations from ACI 318-77 and CP110 which yielded slightly higher values for a low q_e. It was recommended that non-prestressed bonded reinforcement should have been provided when q_e is less than about 0.11, otherwise the flexural instability can result in smaller ductility at failure. The difference of behaviors between the bonded and unbonded slabs with similar reinforcement, and material and loading configurations were found to be very small except the slabs with a very low q_e. Cooke et al. (1981) also observed that the crack spacings of Slab4 were larger than those of its bonded counterpart, SlabB4. A conclusion was drawn that Slab6 and SlabB6 with a very low q_e behaved more like a hinged tied arch than a flexural member. Some of the specimens in these two studies were modeled by the finite element method (FEM) for further study. The descriptions of modeling are presented in Chapter 4.

2.3 Finite Element Formulation of Prestressed Concrete Members

Finite element applications of prestressed concrete (PC) have been actively researched for decades since the pioneering work was done by Ngo and Scordelis (Ngo and Scordelis, 1967). The key to simulating the different types of prestressed concrete systems (pre-tension, bonded and unbonded PT) lies in the modeling of the bonding condition between the concrete and tendons. Tendons in a pre-tensioned prestressed concrete member can be idealized as perfect bonding which implies the strain compatibility between the tendons and surrounding concrete. On the other hand, either a bonded or an unbonded PT member requires an unbonded formulation in the jacking stage. Studies regarding the formulations of the bonded, partially and fully unbonded tendons are various and rich in the literature. The following review discusses the previously developed modeling scheme of interactions between prestressing tendons and concrete. The most popular treatment was to utilize link (rigid spring) elements between tendons and correspond sheathing or concrete. The other method was to employ empiric equations for determining the unbonded tendon strain. However, research on contact techniques to directly model the interface between prestressing tendons and corresponding sheathing has not been conducted to the author's knowledge.

Kang and Scordelis (1980) proposed a nonlinear finite element procedure analyzing prestressed concrete frames with bonded or unbonded tendons. This procedure considers the material and geometric nonlinearities as well as load history, temperature history, creep, shrinkage, and aging of the concrete, and relaxation of the prestress. The complete uniaxial stress-strain relations for the concrete, non-prestressed steel reinforcement and prestressing steel were modeled. The displacement field of the frame element is linear along its local axis and cubic perpendicular to the axis. Reinforcement steel was modeled by a separate layer with the assumption of perfect bonding. The curved prestressing tendons were discretized into several linear segments. The bonded prestress tendons were treated in the same way as non-prestressed bonded reinforcement steel with perfect bonding of the concrete, whereas the strain of the unbonded prestressing tendons was determined by the deformed geometry of the tendons. The prestress loss due to friction and anchorage slip of the PT members at a transfer stage were calculated along with empiric equations and incorporated into the procedure. For the pre-tensioned or PT bonded members, the stiffness of the prestressing tendon was included in the stage of perfect bonding in the analysis, while the stiffness of the prestressing tendon was neglected for the PT unbonded members. The equilibrium equations were formulated as incremental forms of the load. An iterative procedure was applied to solve each load increment. The time domain was divided into a discrete number of time steps to perform a time dependent nonlinear analysis. The aforementioned iterative procedure was applied to each time step to converge the unbalanced load to a prescribed tolerance. This nonlinear finite element procedure was applied to several numerical examples including a bonded PT beam tested by Lin (1955), a bonded and unbonded PT I-beam tested by Breckenridge and Bugg (1964) as well as a pre-tensioned concrete column tested by Aroni (1968). All simulations resulted in good agreement with the experiments.

The finite element procedure proposed by Van Greunen and Scordelis (1983) was able to analyze both the pre-tensioned and PT bonded and unbonded slabs. The procedure was incorporated into a computer program, NOPARC (Van Greunen, 1979) which can analyze the prestressed slabs with the consideration of time dependent effects, material and geometric nonlinearities. A flat triangular shell finite element was used for concrete. The biaxial stress-strain law for concrete was an orthotropic formulation proposed by Darwin and Pecknold (1976; 1977) and adopted in the study. A uniaxial stress-strain relation was used for the non-prestressed reinforcement steel as well as prestressing tendons. The stiffness of the bonded reinforcement steel and tendon was included in the formulation where the deformation field of the steel is the same as that of the concrete. For the unbonded tendons, a modified method utilizing an average extension factor was used to determine the stress-strain state of the unbonded tendons. A step-by-step integration scheme in the time domain was employed to analyze the effects of the time-dependent phenomena such as creep, shrinkage in the concrete, stress relaxation of the prestressing tendons, and temperature changes on the behavior of reinforced and prestressed concrete structures. Two numerical analyses were carried out in the study including a pre-tensioned prestressed concrete column tested by Aroni (1968) and a continuous two-way unbonded PT slab tested by Scordelis et al. (1959). The analysis maintained a good agreement with the experiment for the pre-tensioned column, while the modeling scheme was partially consistent with the slab test data in the inelastic range. Van Greunen and Scordelis (1983) explained that the discrepancies found at the failure stage were caused by the coarse mesh and difficulty of modeling the actual tendon forces at major interior cracks.

El-Mezaini and Citipitioglu (1991) presented a discrete formulation for reinforcement simulating the bonded, fully unbonded and partially unbonded behaviors of the tendons. The key feature of the formulation hinges on the development of an isoparametric element with moveable nodes in which the nodes in the element can be moved without node mapping distortions. With these features, the concrete nodes were moved to new locations, such that the concrete nodes coincide with the reinforcement nodes in space. The stiffness matrices of the concrete and reinforcement elements were evaluated separately. The stiffness matrix of the reinforcement was directly added to that of the concrete for the bonded formulation, whereas the stiffness matrices of the concrete and reinforcement elements were transformed to the local rotated axes per each element. The structural stiffness matrix was reassembled by considering the coupled and uncoupled degrees of freedom of the reinforcement nodes. Linear springs were added between the overlapped concrete and reinforcement nodes simulating a partially bonding condition. Several numerical problems, including a square-plate, a simple-beam with different tendon bonding conditions, a two-span continuous beam as well as a non-prismatic continuous beam with both the bonded and unbonded conditions, were solved by the linear finite element analyses to evaluate the discrete reinforcement formulation. The numerical results yielded good agreement with the previous analysis results by other researchers.

Nikolic and Mihanovic (1996) proposed a finite element model to analyze the PT structures. The tendons were formulated by a one-dimensional isoparametric three-node element with a uniaxial stress-strain relation. The concrete was modeled by an eight-node two-dimensional element associated with a concrete material model which combines the elasto-viscoplastic approach with the smeared crack approach. The prestress loss due to friction was considered by considering that a distributed load due to prestress within a tendon consists of the normal and tangential parts with respect to the tendon direction. The normal part of the distributed loads is a function of the curvature of the tendon, whereas the tangential part is the product of normal part and a friction factor. The prestress transfer was simulated by obtaining the equivalent nodal forces on the concrete nodes due to the prestress. The prestressing tendons and non-prestressed reinforcement were embedded into the concrete with perfect bonding. Typically, three phases were used in a simulation including: 1) applying the dead load prior to prestressing; 2) prestressing; and 3) applying the remaining dead and live load. Four numerical analyses were carried out in the study including an unbonded PT beam, a PT beam with different frictional coefficients, prestressing methods and a bonded PT non-prismatic continuous beam. The numerical results were in good agreement with the available experimental data and previous analyses done by other investigators.

The procedure proposed by Vecchio et al. (2006) could simulate the unbonded PT concrete structures. A link element between the unbonded tendon and concrete was developed which consists of two mutually orthogonal springs. The tangential spring along with the tendon was associated with a bond stress model whereas the other spring had an infinitely large stiffness. The bond stress was modeled as a function of the tendon location, and tendon curvature and wobble effect. The friction stress, however, was independent of the tendon slip. The concrete was formulated by a two-dimensional rectangular element with a smeared rotating crack constitutive law. Prestressing tendons were formulated by a truss element, while non-prestressed reinforcement steel was formulated either by a truss element or a smeared reinforcement layer. Typically, two analysis steps were involved in the proposed procedure: 1) applying the temperature field to the anchorage by which the anchorage expands and shrinks uniaxially in the direction of tendon prestressing to simulate the tendon prestressing and prestress loss due to the anchorage wedge setting; 2) applying an external load. The frictional loss of the tendon was automatically calculated according to the proposed bond stress formulation. Two numerical examples were investigated. A series of flexural controlled prestressed I-beams with unbonded tendons were modeled and had very good agreements with the experiments. Several shear dominant I-beam and T-beams with bonded or unbonded tendons were also modeled, where all of the simulations were well performed compared with the experiments. Vecchio et al. (2006) also concluded that the accuracy of the finite element analysis with respect to the PT unbonded members was barely influenced by neglecting the unbonded tendon friction effect; however, the ductility was greatly impacted during an analysis of a flexural controlled member.

CHAPTER THREE

HYPOTHESIS AND OBJECTIVES

The principal aim of this study is to discuss approaches of modeling a PT prestressed concrete system and accordingly develops finite element modeling schemes appropriate for investigating particular engineering problems. It is hypothesized that potential modeling schemes can be developed via the finite element method (FEM) and thereafter yields robust and accurate numerical solutions. Both the numerical results and previous experimental data were used to make recommendations and suggestions related to the current building code based on the proposed FEM. In part one of this study, modeling schemes based on general purpose finite element packages were discussed and presented which is the basis of the following numerical analyses. Engineering problems that are investigated in this study include the assessment of punching shear provisions in ACI 318-19 (2019) and the evaluation of flexural strength equations (for both bonded and unbonded PT members) in ACI 318-19. The corresponding finite element models are constructed in general purpose finite element packages. In part two of this study, a more flexible nonlinear finite element formulation is developed which specializes in solving structural problems in the context of PT prestressed concrete frames.

The objectives of the research in this book are 1) developing potential modeling schemes for PT prestressed structures based on general purpose finite element packages; 2) evaluating the current building code in terms of particular problems; 3) validation of the proposed modeling schemes and studying specific engineering problems through the proposed FEM; and 4) developing a robust and flexible framework of nonlinear finite element analysis that lies particular emphasis on modeling PT prestressed structures.

The following chapters describe the methodology of this study. The discussions of methodology are divided into two parts. Part one contains the materials of modeling schemes based on general purpose finite element packages (Chapter 4) and corresponding numerical studies regarding engineering problems (Chapter 5). Part two starts at Chapter 6, and mainly discusses the proposed formulations, implementations and mathematic fundamentals that lay behind them. Several numerical examples are carried out in the end of Chapter 6.

Chapter Four

Modeling Scheme in General Purpose Finite Element Packages

4.1 Introduction

This chapter presents and discusses a general methodology of modeling PT prestressed structures in general purpose finite element packages. The specified modeling approaches via ABAQUS (2008) are particularly discussed and all formulations selected are widely available in many other finite element packages. The fundamental principle is the same, even though modeling details might vary as different finite element packages are used. The essential concept is to discretize a concrete entity via general purpose solid elements in combination with truss or beam elements which are used to discretize steel reinforcement.

Constitutive models are carefully selected to address governing laws of materials. Generally, a plasticity model considering strain hardening, strain softening with the ability of adding tension stiffening, proper yielding criterion and flow rules can be successfully used to simulate concrete material with a smeared cracking feature. In some packages, a concrete model with a smeared cracking approach is directly available. Neither type of constitutive model will be able to track individual cracks as cracking is smeared through the entire element. However, it is still possible to capture an individual dominant crack via smeared cracking type model if the mesh is refined in the area of cracking.

A conventional elasto-perfect plasticity model is good enough to model mild steel bars in concrete structures whereas an elasto-plasticity with isotropic hardening is more appropriate of modeling high-strength prestressing steel. Assuming perfect bonding for non-prestressed mild steel bars is accurate enough in most scenarios, since most typical PT structures present very low reinforcement ratios of mild steel bars. On the other hand, interaction between concrete and tendons plays a more significant role than the bonding of mild steel bars. Modeling of unbonded and partially bonded tendons is difficult due to the introduction of a nonlinear boundary. Two modeling schemes (a spring system method vs. a direct contact formulation) are proposed by either approximating the nonlinear boundary problem to a linear boundary problem or formulating the nonlinear boundary via FEM directly. A thorough discussion regarding the spring system method and direct contact formulation is given in this chapter. The prestressing force in a structure is simulated by giving a prescribed prestress field. The initial strain filed is achieved by various methods such as the temperature approach (manipulating the temperature field around tendons).

Conventional implicit static analysis or explicit dynamic analysis is successfully applied to obtain the solution. The choice highly depends on the availability of the algorithm, solution convergence rate and computational efficiency of the selected finite element packages. The aforementioned details are discussed in detail in the last section of this chapter.

4.2 Material Constitutive Relations

4.2.1 Concrete constitutive model

A built-in "damaged plasticity model" in ABAQUS (2008) was employed for concrete modeling. No considerations were given to the concrete damage during unloading, that is, a plasticity model without damage. Preliminary analyses were performed to validate that results under monotonic loads are not sensitive to the damage factors (d_c and d_t in Figure 4.1; zero vs. non-zero). Most documented tests reviewed in the literatures were not subjected to cyclical loads. Therefore, the damage and stiffness degradation descriptions during cyclic loading conditions were not used in the concrete constitutive model for simplicity. The following discussion briefly describes the theories behind the damaged plasticity model in ABAQUS. More detailed information can be found in the ABAQUS theory manual (2008).

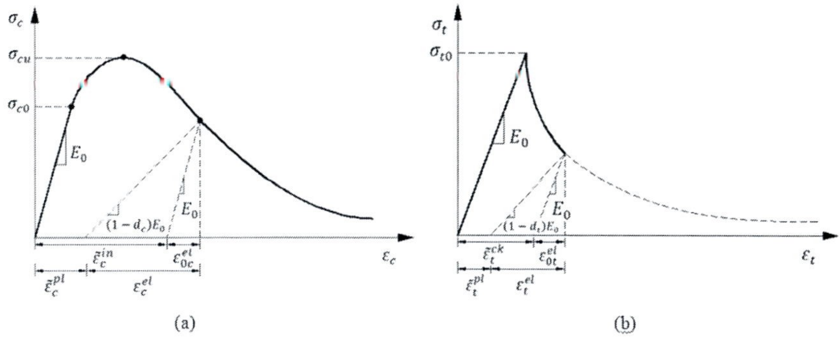

Figure 4.1 Concrete uniaxial stress-strain relation models; (a) in compression and (b) in tension

The expression of strain rate is broken down into elastic and plastic parts as shown in Equation. 4.1.

$$\dot{\varepsilon} = \dot{\varepsilon}^{el} + \dot{\varepsilon}^{pl} \tag{4.1}$$

where $\dot{\varepsilon}$, $\dot{\varepsilon}^{el}$, and $\dot{\varepsilon}^{pl}$ are the total, elastic, and plastic strain rates, respectively.

The strain-stress relations of concrete are presented as:

$$\sigma = (1-d){D_0}^{el}:(\varepsilon - \varepsilon^{pl}) = D^{el}:(\varepsilon - \varepsilon^{pl}) \tag{4.2}$$

where ${D_0}^{el}$ and D^{el} are the initial (undamaged) and degraded elastic stiffnesses of the material, respectively. The parameters of ε and ε^{pl} are the total and plastic strains, respectively (see Figure 4.1), and d is the scalar stiffness degradation variable (= 0 in this study), which is a function of stress state, uniaxial damage variables of d_c and d_t, and stiffness recovery factors.

The relationship between the Cauchy stress (σ) and effective stress ($\bar{\sigma}$) is defined through the scalar degradation variable (d) as:

$$\sigma = (1-d)\bar{\sigma} \tag{4.3}$$

In this study, no damage is considered in the model; therefore, the effective stress simply becomes the Cauchy stress from Equation 4.3.

Hardening variables control the evolution of the yield surface and degradation of the elastic stiffness. They are referred to as the equivalent plastic strains ($\tilde{\varepsilon}^{pl}$) in tension and compression, respectively.

$$\tilde{\varepsilon}^{pl} = \begin{bmatrix} \tilde{\varepsilon}_t^{pl} \\ \tilde{\varepsilon}_c^{pl} \end{bmatrix} \tag{4.4}$$

These hardening variables are a function of three principal strain and principal stress components under general multiaxial stress conditions. The effective uniaxial tensile and compressive cohesion stresses, $\bar{\sigma}_t$ and $\bar{\sigma}_c$, are calculated from the hardening variables as:

$$\bar{\sigma}_t = \frac{\sigma_t}{(1 - d_t)} = E_0(\varepsilon_t - \tilde{\varepsilon}_t^{pl}) \tag{4.5}$$

$$\bar{\sigma}_c = \frac{\sigma_c}{(1 - d_c)} = E_0(\varepsilon_c - \tilde{\varepsilon}_c^{pl}) \tag{4.6}$$

where the subscripts *t* and *c* denote tension and compression, respectively, and E_0 is the initial tangent modulus in uniaxial loading.

As described later in this section, the effective uniaxial cohesion stresses influence yield surface. This model requires uniaxial strain-stress relation to be defined by the user, and then ABAQUS (2008) converts the relation into a stress versus plastic strain curve as expressed in Equations 4.5 and 4.6. Figure 4.1 shows the uniaxial stress-strain relation used in this model for compression and tension, respectively.

The uniaxial stress-strain relation is extended to multiaxial stress conditions. The yield function (F) adopted by this model is based on the yield function proposed by Lubliner et al. (1989) and integrates the modifications suggested by Lee and Fenves (1998), taking the following form:

$$F(\bar{\sigma}, \tilde{\varepsilon}^{pl}) = \frac{1}{(1-\alpha)}\Big(\bar{q} - 3\alpha\bar{p} + \beta(\tilde{\varepsilon}^{pl})\Big\langle \bar{\tilde{\sigma}}_{max}\langle\rangle \Big\langle -\bar{\tilde{\sigma}}_{max}\langle\tilde{\bar{\rangle}}_{c_c}{}^{pl}\Big\rangle\Big\rangle \tag{4.7}$$

where $\langle x\rangle = \frac{1}{2}(|x| + x)$ and $\bar{p}$ is the effective hydrostatic pressure defined as:

$$\bar{p} = -\frac{\bar{\sigma}_x + \bar{\sigma}_y + \bar{\sigma}_z}{3} \tag{4.8}$$

and $\bar{q}$ is the Mises equivalent effective stress defined as:

$$\bar{q} = \sqrt{\frac{3}{2}\bar{S}:\bar{S}} \tag{4.9}$$

where $\bar{S}$ is the effective deviatoric stress matrix. The parameter $\beta(\tilde{\varepsilon}^{pl})$ is a function of the effective tensile and compressive cohesion stresses as well as dimensionless material constant α , which can be determined from the initial equibiaxial and uniaxial compressive yield stresses; the coefficient λ only appears if a triaxial stress state is presented, which can be calculated by comparing the yield conditions along the tensile and compressive meridians; and $\bar{\bar{\sigma}}_{max}$ is the algebraically maximum principal stress of the effective stress tensor.

The plastic potential G was assumed "non-associated" in the meridional (p-q) plane. The potential G used in this model is the modified Drucker-Prager hyperbolic function (Drucker and Prager, 1952; ABAQUS, 2008) defined as:

$$G = \sqrt{(e\sigma_{t0}\, tan\, \psi)^2 + \bar{q}^2} - \bar{p}\, tan\, \psi \tag{4.10}$$

where e is the eccentricity defining the rate at which the function approaches the asymptote and ψ is the dilation angle measured in the p-q plane at high confining pressure.

The constitutive model presented above requires concrete uniaxial stress-strain relation. However, the entire stress-strain curve from experiments was not reported. Instead, concrete compressive strength was accessible in the documentations. Therefore, an empirical model proposed by Carreira and Chu (1985) was employed to describe the uniaxial stress-strain relation in compression of concrete if the experimental data were not available. The tension stiffening was used to define the interaction between the concrete and the deformed bars when cracking occurs. For heavily reinforced members, tension stiffening was suggested such that the stress reduces to zero at a total strain of about ten times the strain at cracking (ABAQUS, 2008). For lightly reinforced members like PT structures, however, such a large tension stiffening effect would introduce unreasonable mesh sensitivity because no reinforcement is provided in some of the cracked regions. A small tension stiffening effect is liable to cause severe discontinuity at the material level. Consequently, a solution slowly converges or diverges in implicit analysis. However, this is not a concern in explicit analysis since ABAQUS (2008) utilizes lumped element mass matrices and equations of motion are integrated by the explicit central difference integration rule. As a result, no stiffness matrix was formed with the desire of solving system equations. For that reason, explicit dynamic analysis was the basis of the following modeling schemes. More discussions of implicit static analysis vs. explicit dynamic analysis are given in Section 4.5. A small tension stiffening effect was considered for the following analyses of PT structures, by assuming the cracking strain ($\tilde{\varepsilon}_t^{ck}$) as equal to approximately twice the strain corresponding to σ_{t0}. Other material parameters described previously, were set to the default values, for no experimental data are available in most cases. However, the ABAQUS manual (2008) does not provide any recommendations for dilation angle ψ. A dilation angle of 50° yielded the best results of preliminary analyses of PT structures; thus, a dilation angle of 50° was recommended and adopted in the following analyses of PT structures.

4.2.2 Steel constitutive model

An elasto-perfect plastic model was assumed for modeling of mild steel (Figure 4.2 (a)). The bilinear stress-strain relationship (along with a yield stress of f_y and elastic modulus of E_s) is reasonably accurate to define the behavior of non-prestressed mild steel bars. The values of f_y and E_s used to define the bilinear relationship were usually measured in experimental programs. Thus, these material parameters were directly obtained from corresponding experiments in the later numerical simulations of documented tests. On the other hand, a nonlinear model consisting of multiple isotropic elasto-plastic segments was used for modeling PT tendons (Figure 4.2 (b)). This nonlinear stress-strain relationship follows the empirical stress-strain model developed by Devalapura and Tadros (1992) for Grade 270 seven-wire strands, otherwise corresponding experimental data were used for the numerical analysis.

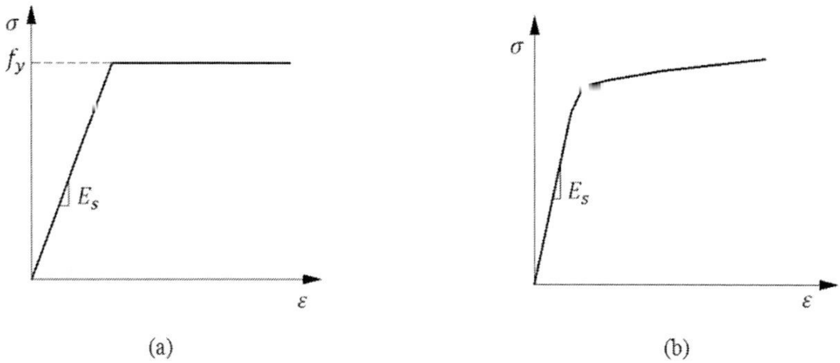

Figure 4.2 Steel uniaxial stress-strain relation models used for: (a) mild steel bars and (b) post-tensioning tendons

4.3 Element Selections

The general purpose 8-node first-order solid element (Figure 4.3) with reduced integration rule (C3D8R in ABAQUS element library) is a good choice for discretizing a concrete entity. This first-order reduced integration element greatly saves computational cost and also avoids shear locking that likely occurs in the first-order full integration element. However, the reduced integration element has some fictitious zero energy modes which lead to overestimated deformation in structures. As a result, a careful hourglass control is imperative for these elements.

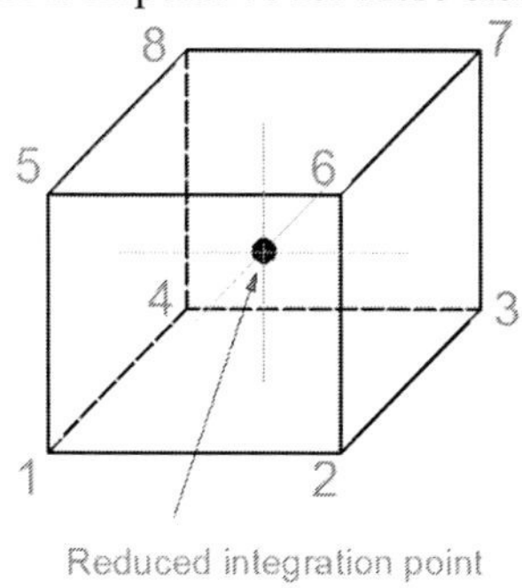

Figure 4.3 Schematic view of the first order brick element with reduced integration rule

Bonded mild steel bars are modeled using 2-node linear truss elements (T3D2 in ABAQUS element library) constrained within the concrete solid elements. The same truss element is applied to model PT tendons despite which interaction model is selected for simulating slips of tendons (spring system method and contact formulation). In both cases, a tendon with a curvy profile is discretized into a large number of linear segments.

If the spring system method is selected to model the interactions of an unbonded PT structure, the spring element needs to be rigid (several magnitudes of order stiffer than other components in the structure) with the ability of finite rotation. The SPRINGA element in the ABAQUS element library is a proper choice for its capability of large displacements in both implicit static and explicit dynamic formulations. The element used for the fictitious tendon is the same as the real tendon. The only discrepancy is that the imaginary tendon has negligible stiffness compared with other components in the structure.

If the contact formulation is selected to model the contact interface, it is necessary to discretize corresponding tendon sheathings. An appropriate choice of elements for modeling tendon sheathings could be the same element used to model concrete entity as a general purpose 8-node first-order solid element. Another choice is a shell element. Although, the element performance of the first-order solid element with reduced integration is very sensitive to the initial configuration of the element shape (a distorted shape of the element will yield inaccurate outputs), it is not a concern to model tendon sheathings which are constrained within concrete elements. The rationale hinges on the typically small stiffness of tendon sheathing compared with the corresponding tendons. Thus, the virtual work contributions from tendon sheathings are negligible. Even if the tendon sheathing has noticeable stiffness along the corresponding tendon direction, it is easier to model by alternative truss elements with equivalent stiffness. Therefore, the tendon sheathing element only serves as a contact interface element which distributes prestressing balancing force into concrete. More details about the spring system method and direct contact formulation are presented in the following section.

4.4 Interaction between Concrete and Steel

4.4.1 Embedding constraint and rigid beam constraint

The interactions between non-prestressed mild steel bars and concrete are assumed to be perfectly bonded as the low reinforcement ratio in typical PT prestressed structures. Based on this assumption, the nodes of mild steel bars need to be displacement compatible with the nodes of surrounding concrete elements. The approaches to assure the compatibility among displacement fields are various, such as superimposing reinforcing bar nodes onto nearby concrete nodes and linear constraint equations. The embedding constraint provided in ABAQUS (2008) is able to constrain translational degrees of freedom which can be used to embed truss elements into solid linear brick elements.

The complicated anchorage of PT prestressing systems can be simplified and modeled by specified nodal constraints. The constraints ensures that the nodes of tendon are properly "anchored" in the concrete at the anchorage zone. The constraint is fulfilled by Multi-Point Constrains (MPCs) provided in ABAQUS (2008). A particular rigid beam MPC is a good choice to anchor the tendon end nodes in surrounding concrete nodes.

4.4.2 Spring system method

The most difficult part of modeling an unbonded or partially bonded PT prestressed system is simulating tendon slips and their corresponding strain. The problem is the nature of boundary nonlinearity. An exact model of the nonlinear boundary requires a contact formulation between tendons and corresponding tendon sheathings. A popular alternative approach to the contact formulation is to approximate the nonlinear boundary to a linear one. It utilizes a large number of linear springs to eliminate the boundary nonlinearity. Figure 4.4 illustrates the mechanism of the spring system model.

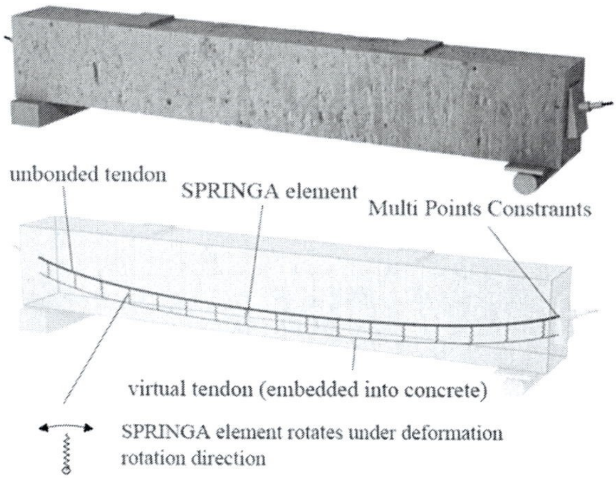

Figure 4.4 Modeling of unbonded PT system using the spring system method

Real and virtual tendons were modeled as depicted, respectively, and they were connected to each other with a spring attached at each node. The springs were assumed to be rigid enough to rotate with no axial and bending deformations. Additional constraints were applied to the springs to ensure that rotation only occurs along the slide direction by linear constraint equations. Virtual (trivial) tendons with negligible stiffness (very small Young's modulus) were embedded in the concrete entity, ensuring no influence on the concrete slab stiffness. As a result, the prestressing balancing load due to post-tensioning was first transferred from the nodes of real tendons to the nodes of virtual tendons, and then to the concrete nodes around virtual tendons while the axial pre-compressive force was transferred by the anchorage simulated by a rigid beam MPC (Figure 4.5). In this manner, the mechanism of unbonded post-tensioning tendons is simulated reasonably well. The discrepancy of prestressing balancing load transfer due to the difference in the position of real and virtual tendons is negligible.

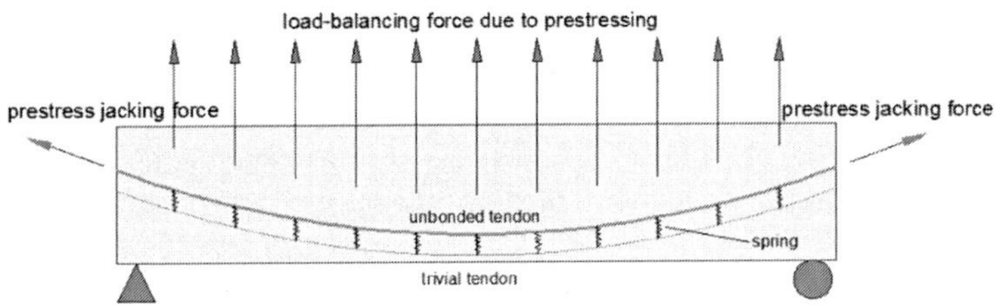

Figure 4.5 Balancing load transferred through spring system

Two questions arise with the spring system method: 1) the optimal spring quantity and location in the modeling, and 2) the optimal length of spring element. The mechanism expressed in Figure 4.3 implies that the primary role of the springs is to transfer the balancing load due to post-tensioning from tendons to the concrete. Thus, the springs should be located at turning points of the tendon and at points where concentrated loads are imposed. To investigate the questions, a parametric study was carried out. A set of analyses was performed with various spring densities for Specimen S1 tested by Foutch et al. (1990), as shown in Figure 4.6 (more information about the modeling and experimental programs is presented in Chapter 5).

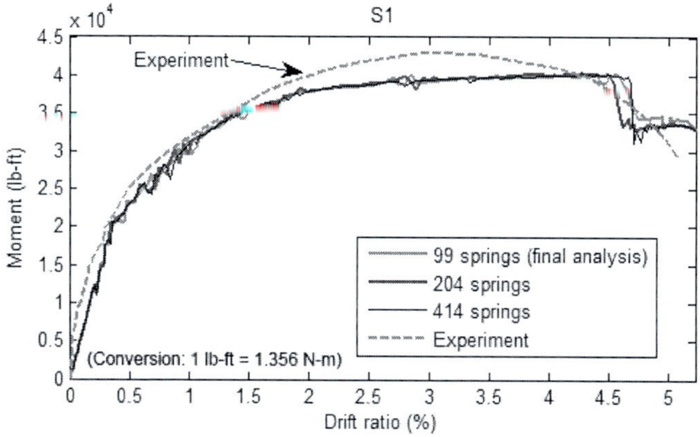

Figure 4.6 Comparisons between the analyses with different quantities of springs used for S1 (Experiments tested by Foutch et al., 1990)

It is evident that the quantity of springs hardly influences the final results and thus, it is concluded that if the springs are located in necessary locations with a reasonable density (e.g., 2 to 3 springs per feet), the results would converge to a unique solution.

Spring length plays another unknown factor in the model. As described, finite sliding was explicitly simulated by the rotation of the spring elements. A smaller radius of rotation may cause a larger vertical position change of the tendon which may in turn lead to a further deviation from the actual solution, whereas a larger radius may too greatly offset the vertical balancing load that is transferred from the real tendon to virtual tendon. Another set of analyses with various spring lengths were made for the same Specimen S1 as shown in Figure 4.7. Three cases were considered: 0.25 in., 0.5 in., and 0.75 in. length for the N-S tendon and 0.5 in., 1 in., and 1.5 in. length for E-W tendon, respectively. Except at the ultimate stage, the responses for the three cases are quite similar. This implies that the spring length in the spring system model is not so influential in the overall results. Rather, the allowable spring length is limited by the concrete cover above or beneath the tendon.

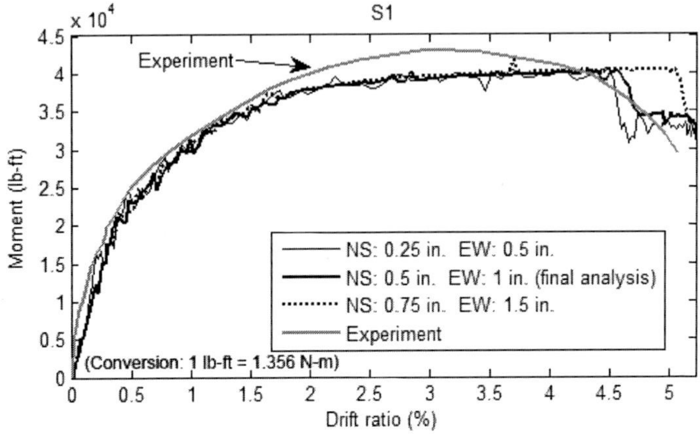

Figure 4.7 Comparisons between the analyses with different spring lengths used for S1 (NS: spring lengths of tendons in N-S direction; EW: spring lengths of tendons in E-W direction; Experiments conducted by Foutch et al., 1990)

4.4.3 Contact formulation

Even though the spring system method is quite flexible in modeling and provides well approximated results, it is limited to modeling perfectly unbonded PT structures which yields a uniformly distributed strain field of tendon. The motivation of investigating the prestressing loss due to friction, simulating partially unbonded and bonded PT structures essentially requires a more robust approach to account for the non-uniform strain field of a tendon. At this point, an appropriate contact formulation might be the route to the ultimate approach of modeling PT prestressed systems. The nature of contact formulation enables the possibility of applying different contact constitutive models at contact interfaces which makes it even more flexible in modeling than the spring system one. The following descriptions of the contact formulation in ABAQUS (2008) are on the basis of explicit dynamic analysis. The interface modeling between tendon elements and corresponding sheathing elements is accomplished through Surface-to-surface contact in ABAQUS/Explicit contact formulations. The tangential behavior of contact is frictionless so that the tendon is free to slip at prestressing stage. The assumption of no frictional effects for unbonded tendon is reasonable (Vecchio et al., 2006; Kang and Wallace, 2008). A simple change of frictionless to rough (infinite friction) ensures the perfect bonding between tendons and corresponding sheathings after prestressing. Sheathings are therefore embedded into the concrete element for deformation compatibility. A similar method based on contact formulation was employed in a numerical study of PT slab in fire by Ellobody and Bailey (2008). Figure 4.8 illustrates the modeling details.

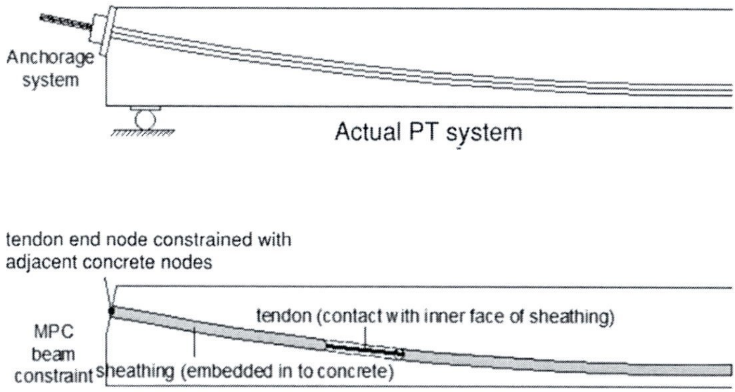

Figure 4.8 Modeling of unbonded/bonded PT systems using the contact formulation

The downside of the current contact formulation compared with the spring system method is its computational anxiety. In the explicit dynamic algorithm, there has not been any strict rule to determine the increment of critical stable time if the contact comes into action. A careful check of system energy is required to validate the solution. Therefore, as a summary, the spring system method offers a quick and simple way to model unbonded PT systems. On the other hand, the contact formulation presents more modeling flexibility and robustness and thereafter is the basis of the following numerical studies in Chapter 5.

4.5 Prestressing and Time Solution Method

The procedure of post-tensioning is complicated as it is accomplished by several stages in real practice. The effective prestress in a tendon can never be uniformly distributed. Even if greased sheathings are used in the prestressing process, prestress loss after post-tensioning is inevitable. However, the magnitude of prestress loss might be very small in those isolated small scale structural members such as in the many down-scaled test specimens reviewed in Chapter 2. This is the basis of assuming a uniformly distributed prestress field immediately after post-tensioning.

This assumption simplifies the implementation of prestressing in the modeling. The popular approach of modeling prestressing is imposing a prescribed prestress field in the tendons before simulations. Due to the nature of explicit dynamic algorithm, to avoid transient response of impulse load, the prestress should be imposed in a gradual manner. Prestressing is obtained by manipulating the temperature field around tendon elements (Stavroulaki et al., 1997) in the case of this study. Uniformly distributed initial prestress is generated by decreasing the temperature field. This approach is reasonable if the length of the tendon is relatively short at the jacking stage where friction loss is a minor effect. All simulations should have a separate step for prestressing in which tendon-sheathing contact is always frictionless. This treatment enables simulation of both unbonded and bonded PT structures. If the unbonded tendon is modeled, the frictionless nature remains in the following analysis steps; otherwise it is changed to rough/stick (infinite friction) simulating bonded PT structures. As mentioned early in this chapter, explicit time integration was employed because the small tension stiffening effect often leads to failures of numerical algorithms in conventional implicit static analysis. On the contrary the explicit algorithm seldom aborts due to its significant robustness. The compromise is the conditional stability of the explicit algorithm which limits the critical time increment to a tiny scale. Even though the time increment is conservatively evaluated in ABAQUS, a careful check of system energy is essential to ensure the solution is not erroneous. In addition, if simulations are performed in a quasi-static manner, a gradually increased load with sufficient period should be able to minimize the inertia effect and damp out transient responses.

CHAPTER FIVE

NUMERICAL SIMULATIONS OF DOCUMENTED TESTS

5.1 Introduction

This chapter presents all numerical simulations that were carried out following the modeling schemes proposed in the previous chapter. These simulations cover four two-way unbonded PT edge slab-column connections tested by Foutch et al. (1990) in Section 5.2; two two-way unbonded PT corner slab-column connections tested by Martinez-Cruzado (1993) in Section 5.3; a two-way bonded PT interior slab-column connection tested by Prawatwong et al. (2007) in Section 5.4; three one-way unbonded PT slabs and three one-way bonded PT slabs tested by Cooke et al. (1981) in Section 5.5; and three PT beams tested by Mattock et al. (1971) in Section 5.5. All of the numerical simulations conducted in this chapter are used to study particular engineering problems in practice in accordance with the current building code (ACI 318-19, 2019). Through Sections 5.2 to 5.4, the complicated moment shear interaction at edge, corner and interior slab-column connections are particularly investigated and studied, respectively. The comparative study of PT tendon bonding and flexural strength are discussed in Sections 5.4 through 5.7. Furthermore, each numerical simulation is compared with corresponding experimental data to validate the modeling schemes proposed in the previous chapter. All modeling approaches are consistent on the basis of discussions in Chapter 4, otherwise will be specified in the modeling descriptions of each section below.

5.2 Numerical Simulations of Four Two-Way Unbonded PT Edge Slab-Column Connections

5.2.1 Description of specimens

Four post-tensioned slab-edge column connections tested by Foutch et al. (1990) were selected to evaluate the developed unbonded PT modeling schemes. Test specimen dimensions and reinforcing details are depicted in Figure 5.1 through 5.3. The half-story column above and below the slab was pinned at each end. In the first two specimens (S1 and S2), tendons were banded perpendicular to the exterior edge of the slab. In the other two specimens (S3 and S4), tendons were banded parallel to the exterior edge of the slab, as presented by Figure 5.1 and 5.2. No. 3 mild steel bars were used as crack-control steel reinforcement (top mild steel) in the vicinity of the column, and were also placed as top and bottom edge reinforcement around the perimeter of the slab to prevent splitting due to PT anchorage bursting forces (see dashed lines in Figures 5.1 and 5.2).

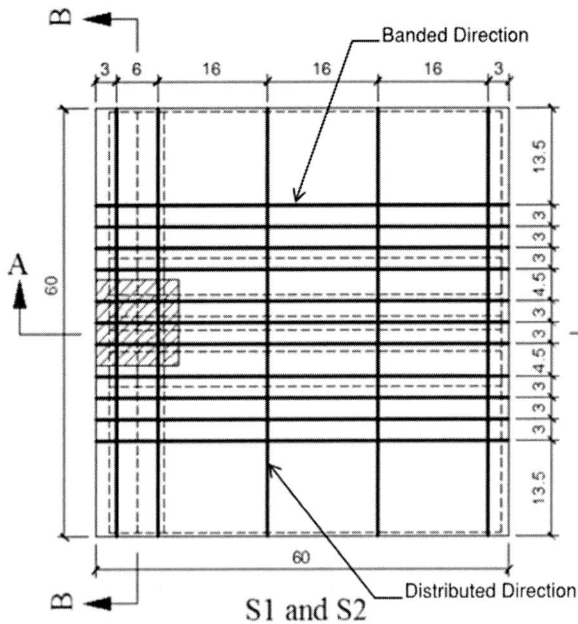

Figure 5.1 Details of reinforcement and dimension of Specimens S1 and S2

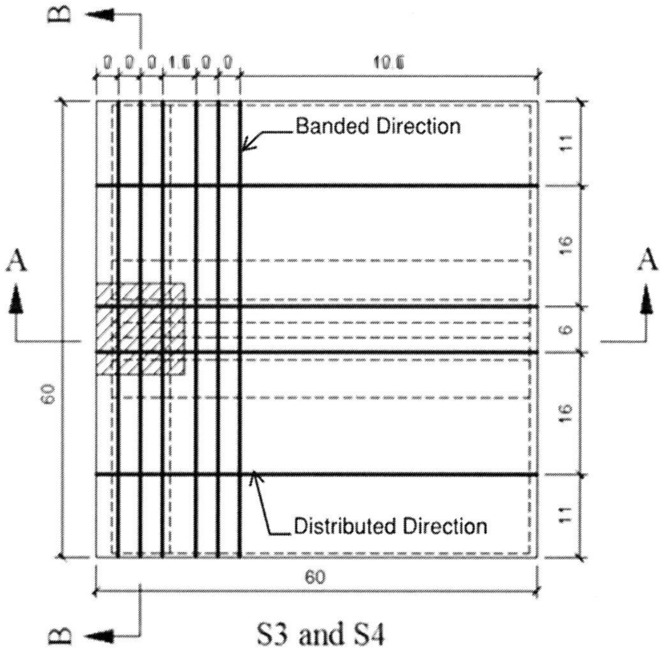

Figure 5.2 Details of reinforcement and dimension of Specimens S3 and S4

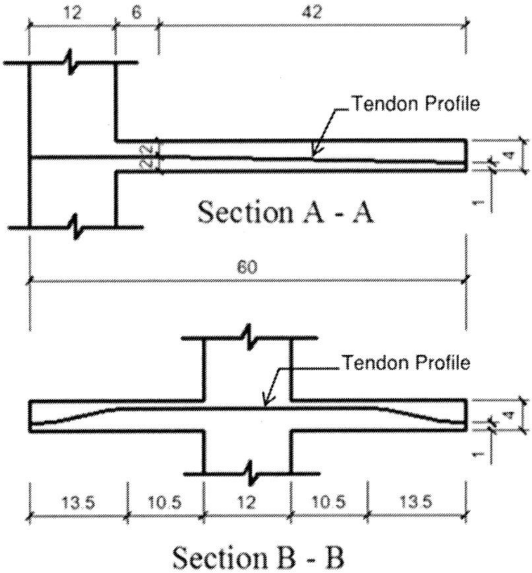

Figure 5.3 Tendon profile in two directions

The only discrepancy between S1 and S2 or between S3 and S4 is the loading position (Figure 5.4), which varies the moment-to-shear ratio. For each specimen, four equally distributed loading plates were placed parallel to the exterior edge of the slab, with the column pinned at both ends (top and bottom). All PT tendons used were Grade 270 3/8 in. diameter, seven-wire strands. The average modulus of elasticity was 28,300 ksi. The tendons were inserted into 1/2 in. diameter polyethylene tubing to prevent bonding to the concrete. All bonded bars were Grade 60 steel with a measured yield stress of 72.7 ksi and an ultimate stress of 126.7 ksi. Detailed information is provided in the paper by Foutch et al. (1990).

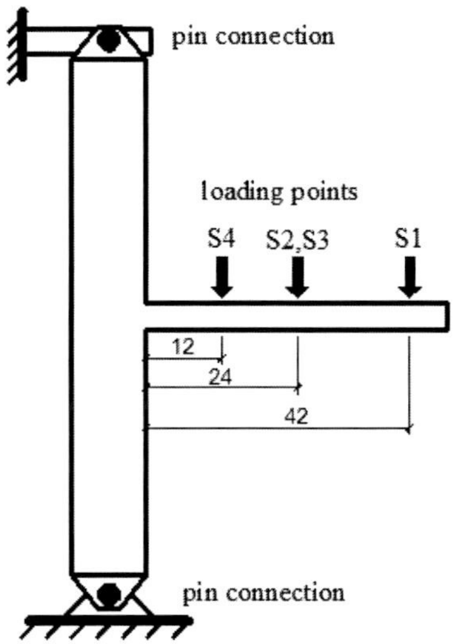

Figure 5.4 Loading positions

5.2.2 Numerical models

The spring system method is adopted for the PT system modeling in this numerical study. One-half of the slab and column was modeled due to its symmetry with respect to the axis perpendicular to the slab edge. The tendon located in the symmetry line was reduced to half the cross-section of the original dimension for modeling symmetrical boundary conditions. For each specimen, about 75 to 100 spring elements were used to simulate the unbonded tendon behavior. Figure 5.5 illustrates the finite element meshes and unbonded tendon modeling systems used for the Specimens S3 and S4 (actual tendons in black and trivial tendons in blue). Every dot except for tendon end anchorages represents the location of a spring element.

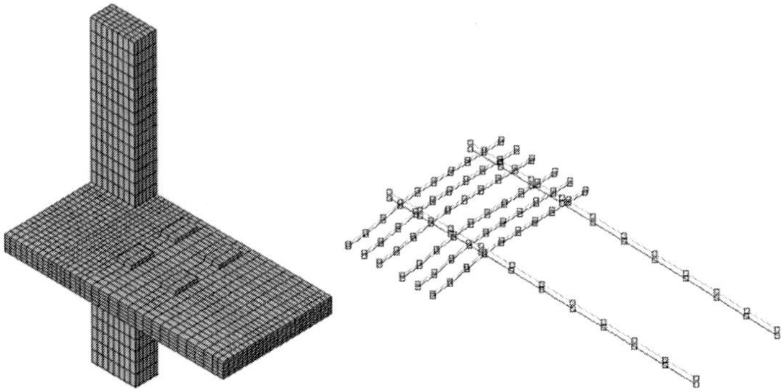

Figure 5.5 Concrete finite element meshes and spring system (half of one PT specimen; symmetric to the centerline of the column)

5.2.3 Numerical results and validations

Figure 5.6 depicts the comparison of the simulated and measured moment versus drift ratio relations for S1 and S3 that failed in a flexural manner. The numerical results of S1 and S3 showed a fairly good agreement with the experimental results.

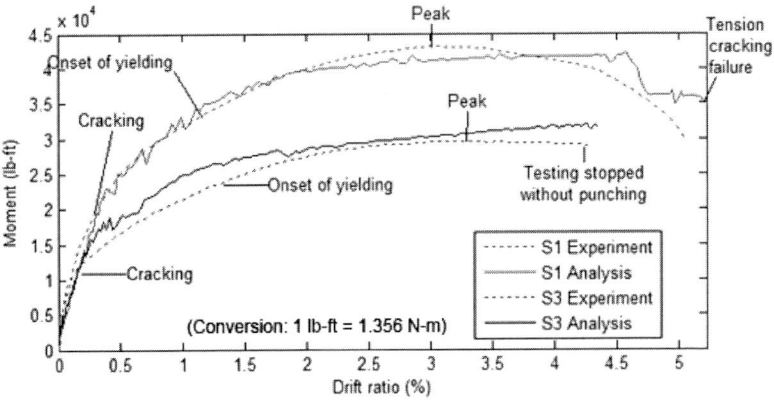

Figure 5.6 Comparisons between experimental and analytical results for S1 and S3

The damage patterns of S1 and S3 are also plotted in Figures 5.7, 5.8 and 5.9. The plot of damage is essentially based on the maximum principal plastic strain which can be considered as cracking. The darker the color, the more tension damage was present in that area. (The same method of post-processing damage pattern was used for the rest of the numerical investigations as in Sections 5.4, 5.4 and 5.5). The experimental result is only available to S1 and is also shown in Figure 5.8. The numerical pattern of S1 agrees very well with the experiment. The dominant crack was especially accurately correctly captured compared with the experiment.

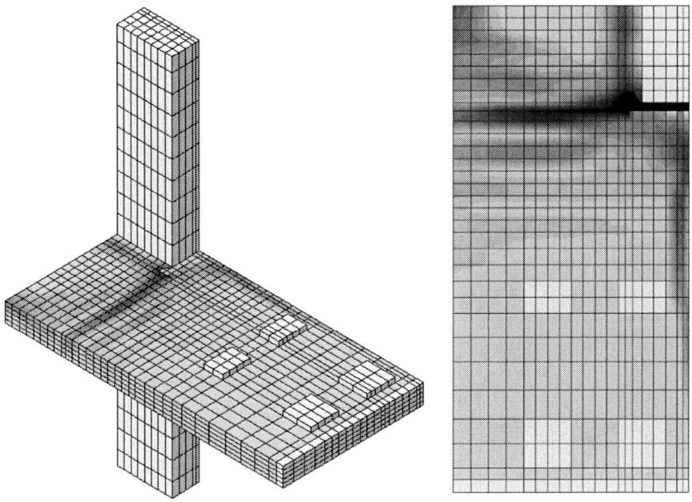

Figure 5.7 Numerical damage pattern of S1 (South half of the original specimen);

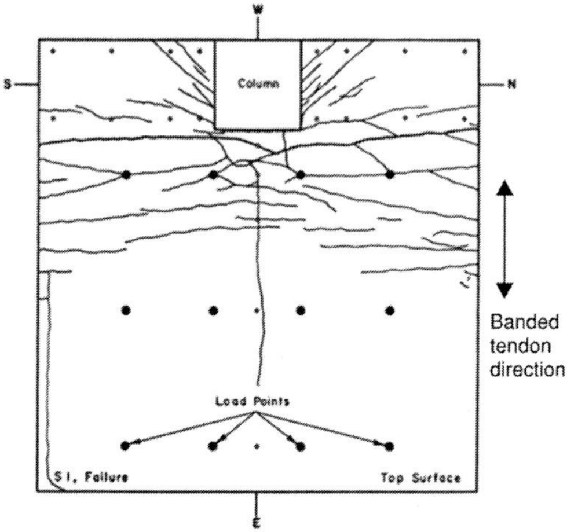

Figure 5.8 Experimental damage pattern of S1 (Foutch et al., 1990)

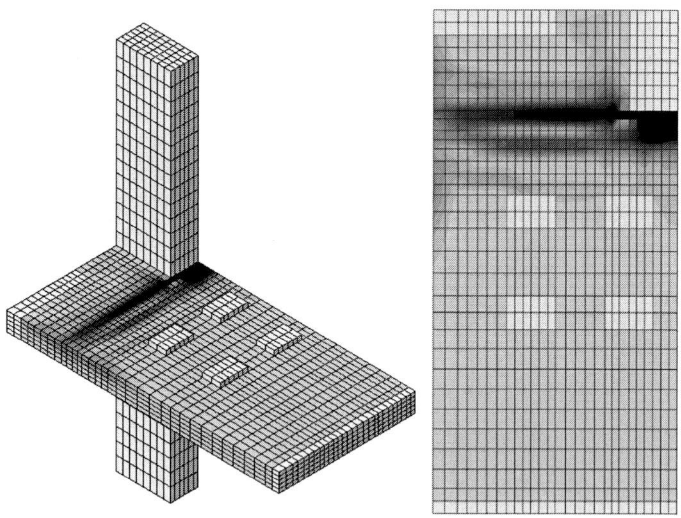

Figure 5.9 Numerical damage pattern of S3 (South half of the original specimen);

Figure 5.10 compares the numerical and experimental results of S2 and S4 that failed in flexure followed by punching. Both S2 and S4 agreed reasonably well with the experimental data prior to punching shear failure. Although a slightly higher strength resulted than that found from the measured moment-drift relationships, general trends correlated fairly well with the test data. Furthermore, the PT connection damage patterns observed by Foutch et al. (1990) correspond to the numerically predicted behavior (Huang et al., 2010). According to the deformations resulting from the analysis, S1 and S3 were simulated to experience very large slab rotations at the column face, whereas S2 and S4 were simulated to produce significant shear deformations near the column face. These are consistent with the test observations.

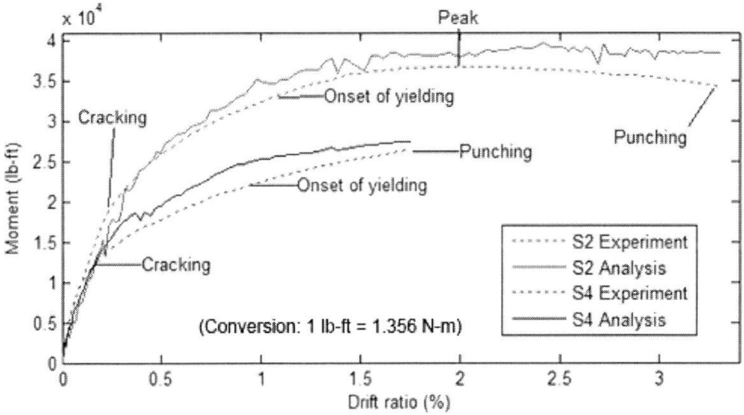

Figure 5.10 Comparisons between experimental and analytical results for S2 and S4

The damage patterns from numerical simulations and experiments are presented in Figures 5.11, 5.12 and 5.13 for S2 and S4, respectively. The available experimental data for S2 is also shown in Figure 5.12. The comparison between numerical simulation and experiment indicates a good agreement in terms of the location and pattern of cracks.

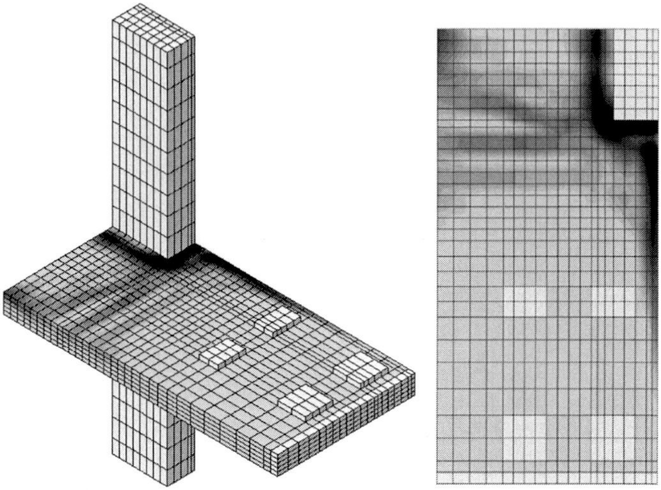

Figure 5.11 Numerical damage pattern of S2 (South half of the original specimen); (a) perspective view and (b) plan view

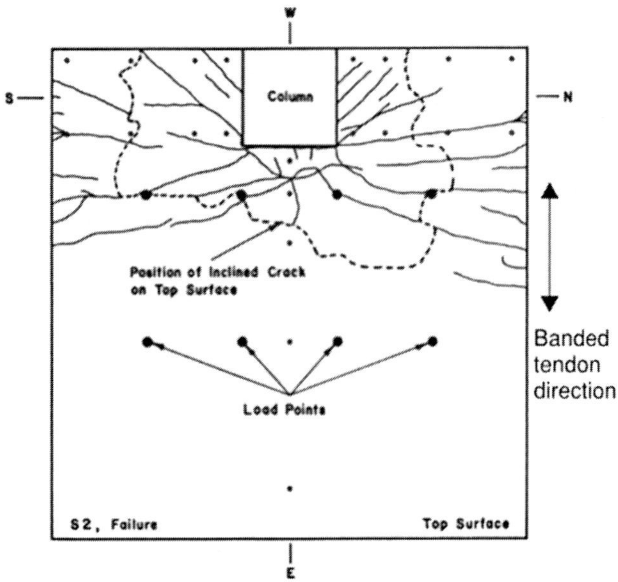

Figure 5.12 Experimental damage pattern of S2 (Foutch et al., 1990)

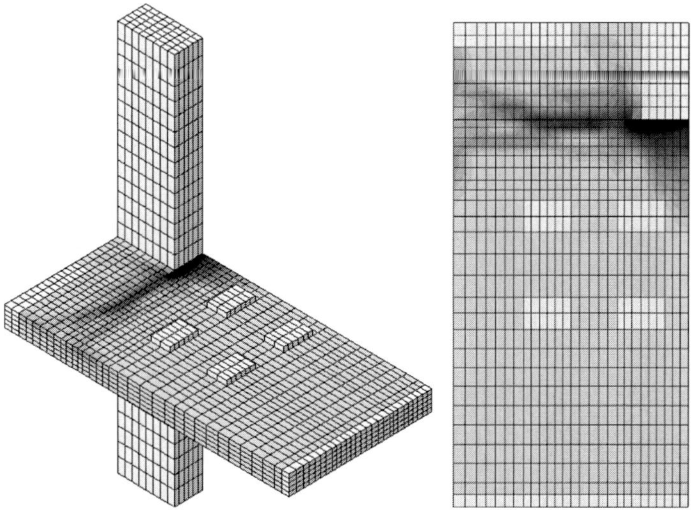

Figure 5.13 Numerical damage pattern of S4 (South half of the original specimen); (a) perspective view and (b) plan view

Tendon stress values at multiple locations in a tendon were evaluated. The uniform tendon stress readings throughout the entire length demonstrate that the strain field of unbonded tendon was properly simulated. Figure 5.14 plots the maximum tendon stress development with increasing moment for the specimens. The tendon stress increases were in very good agreement with experimentally monitored increases, validating that the spring system method for unbonded tendon modeling is robust and accurate.

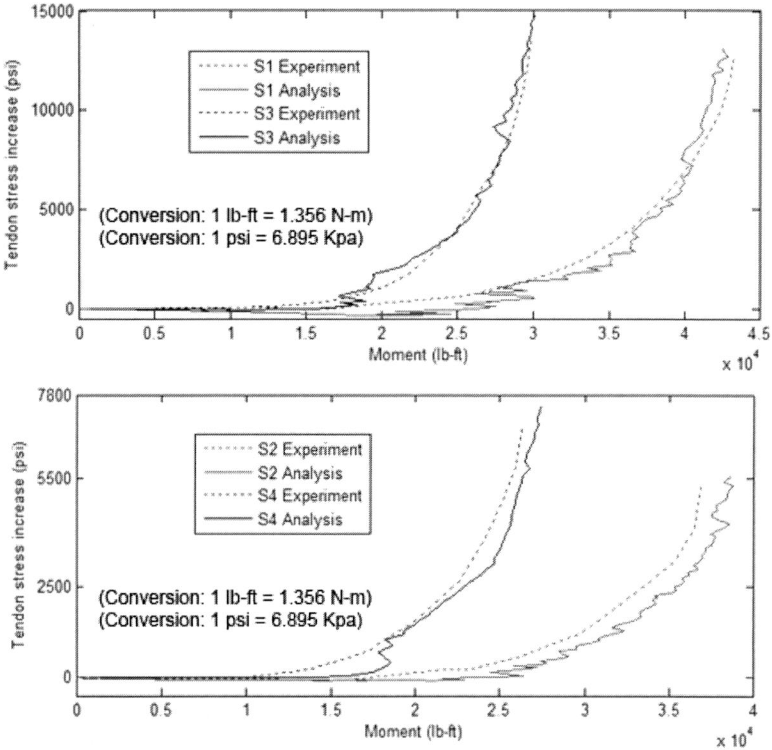

Figure 5.14 Unbonded tendon stress increases versus applied moment

5.2.4 Assessments of ACI 318-19 punching shear provisions

Assessment of γ_v factor

Moment and shear transfers at slab-column connections are described in ACI 318-19, Sections 8.5.3. and 22.6.4 (ACI 318-19, 2019). ACI 318 code provisions present an empirical model, the so-called eccentric shear stress model, for designing shear and moment transfer on a critical section located at ($d/2$) from the column face. The sum of direct shear and eccentric shear due to a fraction of unbalanced moment transfer may cause punching shear failure. A fraction of unbalanced moment given by $\gamma_f M_u$ is considered to be transferred by flexure, which suggests that the rest of the unbalanced moment, $\gamma_v M_u$, is assumed to be transferred by eccentric shear stress on the critical section. The fraction factors of γ_f and γ_v are given by ACI 318-19, Sections 8.4.2.2.2 and 8.4.4.2, respectively, as:

$$\gamma_f = \frac{1}{1 + (2/3)\sqrt{(b_1/b_2)}} \tag{5.1}$$

$$\gamma_v = (1 - \gamma_f) \tag{5.2}$$

where b_1 is the width of the critical section measured perpendicular to the axis about which the moment acts and b_2 is the width of the critical section transverse to b_1. For reinforced concrete slab-column connections, ACI 318 allows the fraction of the unbalanced moment being transferred in flexure to be increased for both exterior and interior connections with relatively small gravity shear. The fraction factor of γ_f may be increased up to 1.0 for edge connections and 1.25(γ_f) for interior connections, if their factored gravity shear ratio ($V_u/\phi V_c$) is not more than 0.75 and 0.4, respectively (Section 8.4.2.2.4), where V_u is the factored direct shear force to be transferred from the slab to the column, ϕ is the strength reduction factor of 0.75 and V_c is the nominal shear capacity provided by concrete. Also, to apply for this provision, the net tensile strain (ε_t) at ultimate within the effective transfer slab width of an interior connection should not be less than 0.01. This provision gives great flexibility in connection design. However, ACI 318-19 does not permit adjustments of γ_f or γ_v for prestressed (post-tensioned) connections. This calibration study aims to assess the validity of the model and provision, and to extend or improve the model, particularly the fraction factor of γ_v defined as part of the model.

Based on the well-developed finite element models, the value of γ_v has been evaluated for each PT specimen. The average shear stress along the slab thickness on a critical section has been monitored from integration points of the 5 layers (elements) located at the front corner of the shear critical section (see Figure 5.15 (a)).

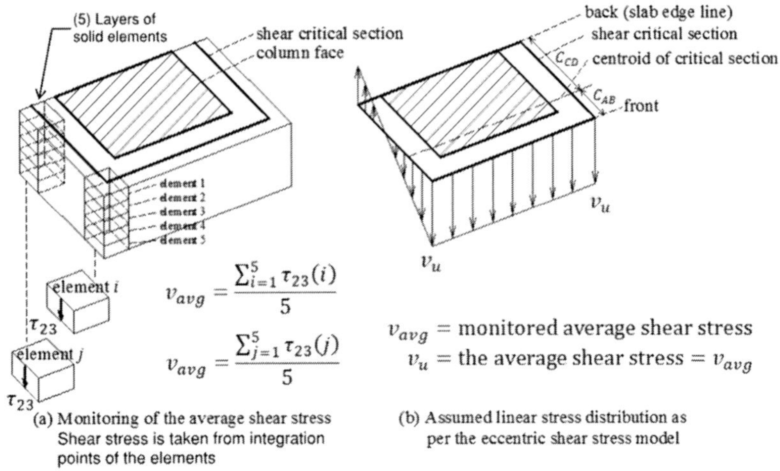

Figure 5.15 Monitoring of direct and eccentric (torsional) shear stresses associated with the eccentric shear stress model

The analysis found that the shear stress on the side face at the front or back corner of the critical section was always the largest. The previous finite element analysis for reinforced concrete exterior slab-column connections (e.g., Park and Choi, 2007) also reported similar results. The shear stress at this location, in accordance with ACI 318-19, is assumed to be equal to direct shear plus eccentric shear due to the fraction of unbalanced moment transfer (see Figure 5.15 (b)). The applied direct shear (V_a) and unbalanced moment ($V_a l'$) were determined by statics, and the average shear stress (v_u) was directly obtained from the finite element analysis (Figure 5.15 (a)), where l' is the distance from the centroid of the critical section to loading point. The direct shear (V_{sw}) and unbalanced moment (M_{sw}) due to self-weight were collected at the location of the column center prior to the monotonic analysis. Finally, the value of γ_v was determined from the following equations:

$$v_u = \frac{(V_{sw} + V_a)}{A_c} + \frac{\gamma_v (M_{sw} - V_{sw} g + V_a l') c_{AB}}{J_c} \tag{5.3}$$

$$v_u = \frac{(V_{sw} + V_a)}{A_c} - \frac{\gamma_v (M_{sw} - V_{sw} g + V_a l') c_{C'D'}}{J_c} \tag{5.4}$$

where A_c is the area of shear critical section specified in ACI 318-19 (2019); g is the distance from the centroid of the critical section to the column center, c_{CD}, (c_{CD} minus one half the element dimension perpendicular to the slab edge at shear critical section = c_{CD} – 1.5 in.) [note that v_u is monitored slightly inside the slab edge and see element j in Figure 5.15 (a)]; c_{AB} and c_{CD} are the distances from the centroid of the critical section to the perimeter of the critical section on the front end and back end (slab edge), respectively (see Figure 5.15 (b)); and J_c is the polar moment of inertia of the critical section (see Park and Gamble, 2000; Section 10.3.3).

Figure 5.16 illustrates values of γ_v monitored at each loading step. Because flexural yielding of the bonded steel occurred prior to (or without) punching failure, which is common in a typical PT design, γ_v values in the inelastic deformation range or at peak strength of these specimens are primarily of interest. Beyond yielding (after 1% drift ratio according to the paper by Foutch et al., 1990), the γ_v factor generally decreased as the load increased (i.e., interaction between moment and shear decreased). The values at peak and punching or tensile cracking failure were less than those specified in ACI 318-19 (see dashed lines in Figure 5.16). For example, the monitored values were only about half the specified value at peak or punching. It is notable that the γ_v values obtained from the front and back sides of the critical section (i.e., from Equations 5.3 and 5.4) are quite consistent, which indicates that the eccentric shear stress model is valid for PT edge slab-column connections and that there is a clear interaction between moment and shear at the PT edge connection, unlike the case of RC edge connections. Only the γ_v values at the front and back sides of S4 are quite different from each other, due to the small applied moment-to-shear ratio (i.e., small l'; see Figure 5.4), which is unlikely to occur in real cases. For small unbalanced moment ($V_a l'$), the results are very sensitive to the variation of V_a.

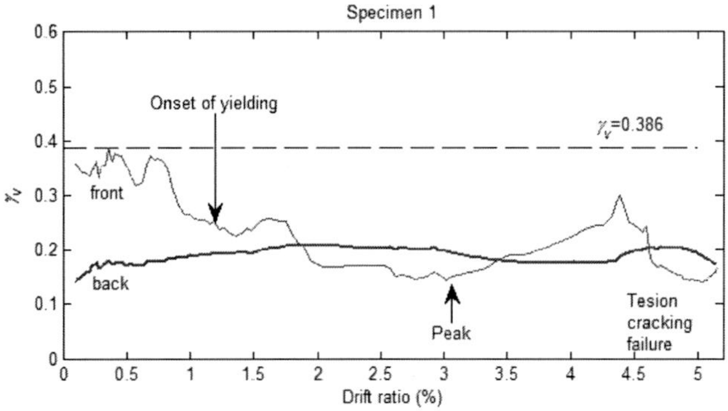
Specimen 1
0.6
0.5
0.4
0.3
0.2
0.1
0
γ_v
Onset of yielding
γ_v=0.386
front
back
Peak
Tesion cracking failure
0 0.5 1 1.5 2 2.5 3 3.5 4 4.5 5
Drift ratio (%)

(a)

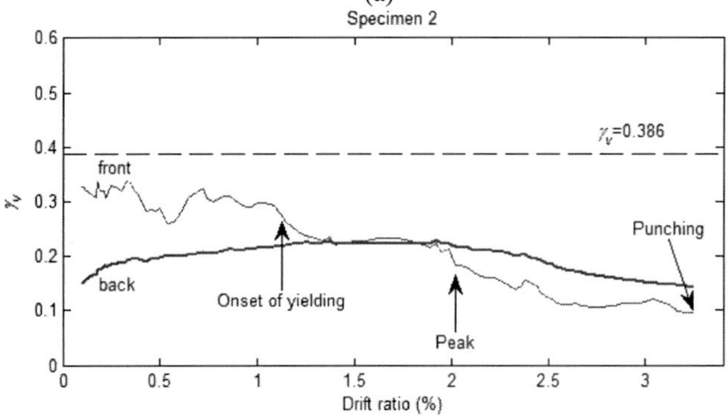
Specimen 2
0.6
0.5
0.4
0.3
0.2
0.1
0
γ_v
γ_v=0.386
front
back
Onset of yielding
Peak
Punching
0 0.5 1 1.5 2 2.5 3
Drift ratio (%)

(b)

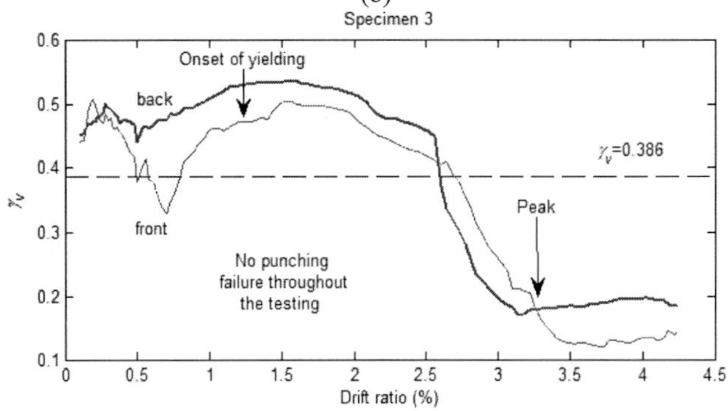
Specimen 3
0.6
0.5
0.4
0.3
0.2
0.1
γ_v
Onset of yielding
back
γ_v=0.386
front
Peak
No punching failure throughout the testing
0 0.5 1 1.5 2 2.5 3 3.5 4 4.5
Drift ratio (%)

(c)

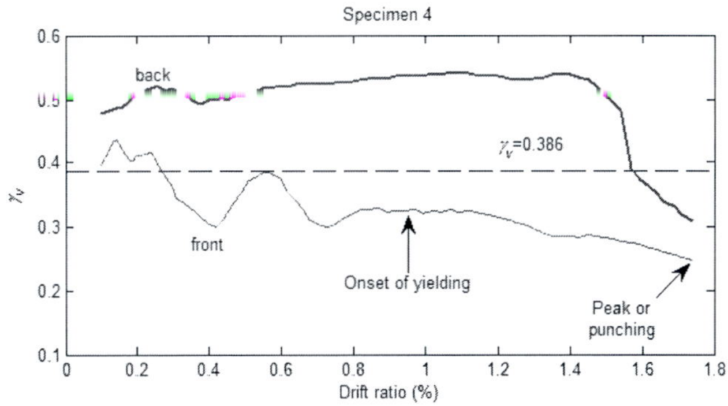

(d)

Figure 5.16 Numerical results of the fraction (γ_v) of unbalanced moment transferred by eccentric shear for edge PT slab-column connections

According to the finite element results, S3 and S4 had quite large fractions (γ_v) of the unbalanced moment transferred by eccentric shear at the corner of the connection before yielding, whereas at peak or punching, the γ_v values of S3 and S4 were substantially decreased. The higher γ_v at lower drifts was related to the tendon arrangement. The shear stress of □23 (see Figure 5.15 for notations) was relatively large before the redistribution of shear stress because of large initial prestress (normal stress) from banded tendons parallel to the slab edge, and it was verified that the vertical shear stress was proportional to the initial normal stress of the element at the location. After the stress redistribution, the simulated behavior was similar to S1 or S2. The tendons placed adjacent to the column produced higher shear stress on the side critical section of S3 than S1, but did not cause serious shear damage, possibly due to the confining effects. A similar behavior was obtained in S4, which failed in punching shear. Note that S4 resisted a maximum shear stress substantially higher than the concrete shear stress capacity with consideration of f_{pc} (dotted lines in Figure 5.16). More discussions on the shear stress capacity are presented in the following section.

Although it may not be possible to make clear recommendations given the limited calibration exercise, application of reduced γ_v (e.g., by 50%) may be allowable for edge PT slab-column connections transferring moment normal to the slab edge. This exercise is highly intriguing, and there is significant potential in further assessment of more PT edge slab-column connections and PT connections with lower precompression in concrete, different aspect ratios and/or various loading and boundary conditions.

Assessment of punching shear stress capacity (v_c)

ACI 318-19, Section 22.6.5.5 defines the punching shear strength of an interior PT slab-column connection as follows:

$$v_c = (\beta_p \lambda \sqrt{f_c'} + 0.3 f_{pc}) b_o d + V_p \tag{5.5}$$

where β_p is the smaller of 3.5 and ($\beta_s d/b_o$ + 1.5, with β_s = 40, 30 and 20 for interior, exterior and corner connections, respectively), if f'_c is in psi, λ is the lightweight concrete modification factor (λ for PT slabs with normal weight concrete = 1), b_o is the perimeter of the assumed critical section, d is the effective depth, f_{pc} is the average compressive stress in concrete due to the effective post-tensioning force for the full specimen width, and V_p is the vertical component of all effective post-tensioning forces crossing the critical section. The V_p term is approximately neglected, as the tendon profile was relatively straight with almost zero eccentricity for the scaled edge connection with a very small tributary slab area. Equation 5.5 is not permitted if the following limits of ACI 318-19 Sections 22.6.5.4 and 22.6.5.5 are not satisfied:

(a) No portion of the column cross section shall be closer to a discontinuous edge than four times the slab thickness;

(b) The value of $\sqrt{f_c'}$ used in Equations 5.3 and 5.4 (in this book) shall not be taken greater than 70 psi; and

(c) In each direction, f_{pc} shall not be less than 125 psi, nor be taken greater than 500 psi.

The specimens tested by Foutch et al. (1990) do not satisfy most of these limits. Primarily, according to the first limit, an edge PT slab-column connection is assumed to have a smaller punching shear strength, which is equivalent to that of an RC slab-column connection without post-tensioning, than Equation 5.5. ACI 318-19 Commentary R22.6.5.4 states that the prestress is not fully effective around the perimeter of the critical section near the slab edge. In this calibration study, investigation is focused on whether the aforementioned statement is the case, and if not, on how much concrete shear stress can be exerted on the critical section prior to punching failure, following flexural failure. Note that most typical PT connections experience flexural yielding followed by punching shear, and that the punching failure without yielding of bonded steel is quite a rare case of design.

Previous experimental research (e.g., Smith and Burns, 1974; Trongtham and Hawkins, 1977; Foutch et al., 1990) attempted to obtain the punching shear capacity of individual PT slab-column connections (without shear reinforcement) by monitoring applied shear and moment (measured). However, actual stresses were not obtained from any of the previous tests due to the difficulty in monitoring the applied shear force or stress at a point inside the concrete slab. Furthermore, there have been significant uncertainties of the effective transfer slab width, the fraction factor of γ_v and the unbonded tendon stress at punching. Well-developed nonlinear finite element modeling, along with documented experimental data, innovatively solves this problem.

Figure 5.17 illustrates the variation of average shear stresses (v_u) at the front and back corner points of the critical section compared to the ACI 318 shear stress capacities of PT interior and exterior connections (v_c for exterior, shown using dots; and v_c for interior, shown using dashed line). As depicted in Figure 5.15 (a), the average of the five layers' shear stresses was directly obtained from the finite element analysis at each loading step. Specimen S1 failed in a ductile manner with a very high moment-to-shear ratio (see Figure 5.4). A flexural yield line was formed across the full width of the slab before the connection exhausted its capacity to transfer the unbalanced moment and shear. A rather ductile mode of tension cracking failure occurred on the top of the slab along the column face due to considerable slab folding. Specimens S3 and S4 achieved higher shear stresses than those from Equation 5.5, and S2 almost reached the concrete stress capacity of $(3.5\sqrt{f_c'} + 0.3f_{pc})$. It is noted that actual shear stress capacity appears to be much larger than the maximum monitored v_u, as the specimens except S4 failed in a ductile manner without or prior to punching. If the specimens (e.g., S1 and S2) would be designed with larger flexural capacity (e.g., by providing additional bonded steel), shear failure would occur at much higher v_u (this has been confirmed from an additional analysis); however, such brittle design is not feasible in practice.

Although it appears that the PT specimens had a concrete shear capacity greater than Equation 5.5, additional studies on other PT edge connections would be needed to develop recommendations on the punching shear stress capacity of a PT edge connection or to relieve the limitation of $\sqrt{f_c'}$ and/or f_{pc} for PT connections.

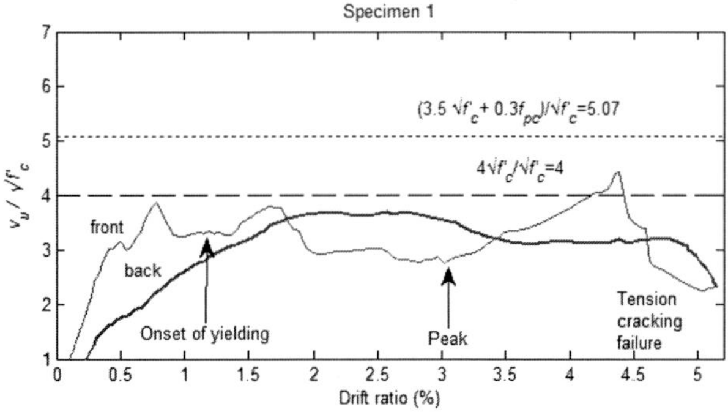
Specimen 1
(3.5 √f'c + 0.3fpc)/√f'c=5.07
4√f'c/√f'c=4
vu / √f'c
front
back
Onset of yielding
Peak
Tension cracking failure
Drift ratio (%)

(a)

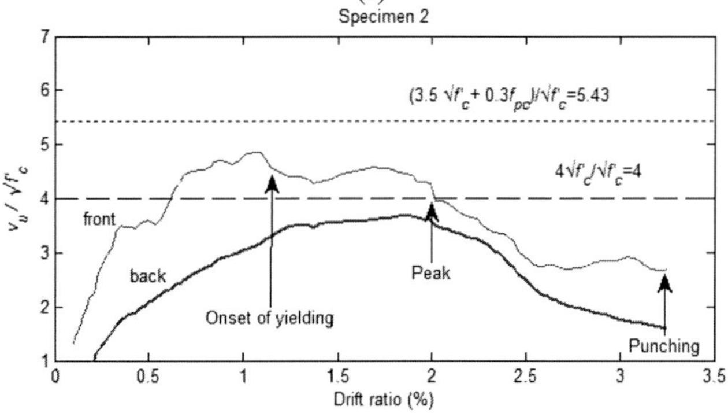
Specimen 2
(3.5 √f'c + 0.3fpc)/√f'c=5.43
4√f'c/√f'c=4
vu / √f'c
front
back
Onset of yielding
Peak
Punching
Drift ratio (%)

(b)

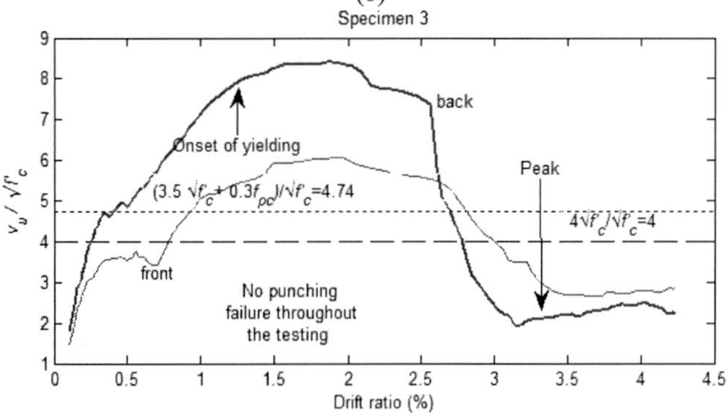
Specimen 3
back
Onset of yielding
(3.5 √f'c + 0.3fpc)/√f'c=4.74
Peak
4√f'c/√f'c=4
vu / √f'c
front
No punching failure throughout the testing
Drift ratio (%)

(c)

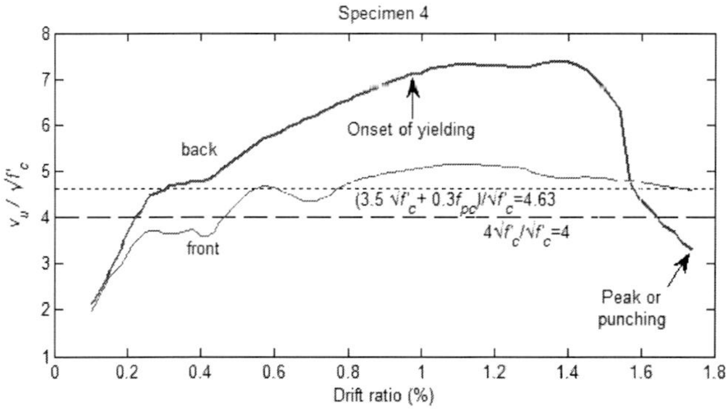

(d)

Figure 5.17 Finite element results of average shear stress (v_u) at the corner of the critical section for edge PT slab-column connections

Assessment of ACI 318 punching shear provisions of PT edge connections

Table 5.1 summarizes the test results of direct shear and unbalanced moment at peaks of applied loads, as well as the values of v_u calculated using these experimentally measured values and Equations 5.3 and 5.4, and the values of V_c based on Equation 5.5 and ACI 318 code provisions. When comparisons are made between the larger of the two values in Columns [11] and [12] of Table 5.1 and the specified capacity in Column [13] (or Column [14] based on Equation 5.5) of Table 5.1, it is apparent that the ACI 318 code provisions are overly conservative. Again, note that under applied direct shear and unbalanced moment, the specimens did not undergo punching prior to flexural failure. Furthermore, the back of the critical section (i.e., slab edge) was observed to be minimally damaged, as opposed to the values indicated in Table 5.1 (see Column [11] vs. Column [12]). If the γ_v factor is reduced to one-half of its specified value, the values would make more sense. This exercise validates the findings obtained from applying nonlinear finite element modeling.

Table 5.1 Shear stress demands and capacities calculated based on test data for PT specimens

Specimens		[1]	S1	S2	S3	S4
V_{sw}	(kips)	[2]	1.45	1.45	1.45	1.45
V_a	(kips)	[3]	12.21	18.4	14.52	26.3
M_{sw}	(in.-kips)	[4]	38.3	38.3	38.3	38.3
A_c	(in^2)	[5]	140.6	140.6	140.6	140.6
l'	(in.)	[6]	44.7	26.7	26.7	14.7
G	(in.)	[7]	3.28	3.28	3.28	3.28
c_{AB}	(in.)	[8]	4.37	4.37	4.37	28,0
c_{CD}	(in.)	[9]	9.28	9.28	9.28	9.28
J_c	(in^4)	[10]	2242.7	2242.7	2242.7	2242.7
v_{u_AB}‡	(psi)	[11]	533.5	533.3	427.6	508.6
v_{u_CD}‡	(psi)	[12]	-767.2	-636.3	-498.4	-413.3
v_c†	(psi)	[13]	341.8	315	312.4	334.7
v_c††	(psi)	[14]	433.4	428	370.1	387.6
v_{u_AB}‡‡	(psi)	[15]	315.3	337.3	270.6	353
v_{u_CD}‡‡	(psi)	[16]	-335	-247.6	-192.4	-107.9

Notations are the same as used in Equations 5.3 and 5.4, except the following:
(1) V_{sw} is the experimentally measured direct shear due to self-weight of the specimen and setup at the centroid of the critical section when V_a was reached (Foutch et al., 1990)
(2) V_a is the peak applied load obtained from the experiment (Foutch et al., 1990)
(3) M_{sw} is the experimentally measured unbalanced moment due to self-weight of the specimen and setup at the centroid of the critical section when V_a was reached (Foutch et al., 1990)
‡ : Shear stress calculated using the values in Table 5.1 and Equations 5.3 and 5.4, with γ_v (= 0.386)
‡‡ : Shear stress calculated using the values in Table 5.1 and Equations 5.3 and (5.4), with γ_v (= 0.5 x 0.386)
† : Shear stress capacity calculated using the values from experiments and ACI 318-19 [= $(4\sqrt{f_c'})$ psi]
†† : Shear stress capacity calculated using the values from experiments and Equation (5.5) [= $(3.5\sqrt{f_c'} + 0.3f_{pc})$ psi]

5.2.5 Conclusion

Realistic modeling of unbonded post-tensioned slab-column connections is particularly challenging due to its complex three-dimensional stress states and non-conventional interaction between unbonded tendons and concrete. In this study, the spring system method along with other proposed modeling schemes have been employed to simulate unbonded post-tensioned connection behavior. Based on the modeling and calibration studies, the following conclusions were drawn:

(1) The accuracy of the unbonded tendon modeling approach has been demonstrated through the direct comparison of overall behavior and damage patterns, moment-drift relations, and unbonded tendon stress increases. The developed nonlinear finite element model performed considerably well for all PT slab-column edge connections with two different tendon layouts (banded-distributed and distributed-banded) and three different moment-to-shear ratios.

(2) The fraction (γ_v) of the unbalanced moment being transferred by eccentric shear was estimated to be about half the ACI specified value at peak or punching for the edge PT slab-column connections. This indicates that there were moderate interactions between moment and shear, and thus a decrease in γ_v by 50% (or increase in γ_f by 50%) may be permitted for an edge PT slab-column connection. However, additional calibration work, perhaps using PT slabs tested by other investigators, would be needed for a thorough investigation on the γ_v factors of PT slab-column connections.

(3) The assessment of the γ_v factors indicates that the eccentric shear stress model is valid for PT edge slab-column connections and that there is a certain interaction between moment and shear at the PT edge connection, even though it is very small.

(4) The punching shear capacity of edge PT slab-column connections appears to be benefitted by the prestress due to post-tensioning, as opposed to the ACI 318 code requirements and commentary (Sections 22.6.5.4 and 22.6.5.5 and R22.6.5.4). This is the case for both banded-distributed and distributed-banded tendon arrangements. However, in order to apply Equation 5.5 for an edge PT connection, additional studies on other PT connections would be needed. Furthermore, the limitation of $\sqrt{f_c'}$ and/or f_{pc} for PT connections could be relieved.

5.3 Numerical Simulations of Two Two-Way Unbonded PT Corner Slab-Column Connections

5.3.1 Description of specimens

Two isolated corner Specimens C1 and C2 tested by Martinez-Cruzado (1993) were selected to evaluate the proposed finite element modeling schemes and ACI 318-19 punching shear provisions (ACI 318-19, 2019). C1 and C2 have identical geometry and reinforcement whereas the only discrepancy between two specimens are different prestressing force, gravity load and test procedures. The following descriptions are based on C1. The prototype building was scaled down to two 3/7 scale isolated corner connections with a termination boundary located at inflection points at the original building. Five 3/8 in. diameter prestressing strands were banded in the north-south direction concentrating at the column strip while another two prestressing strands were banded and three prestressing strands were uniformly distributed outside column strip along the east-west direction. All prestressing strands were inserted into flexible pvc tubes to maintain an unbonded interface between the strand and the concrete. Mild steel bars were provided only at the negative moment region around column. Details of dimensions and reinforcement are depicted in Figure 5.18.

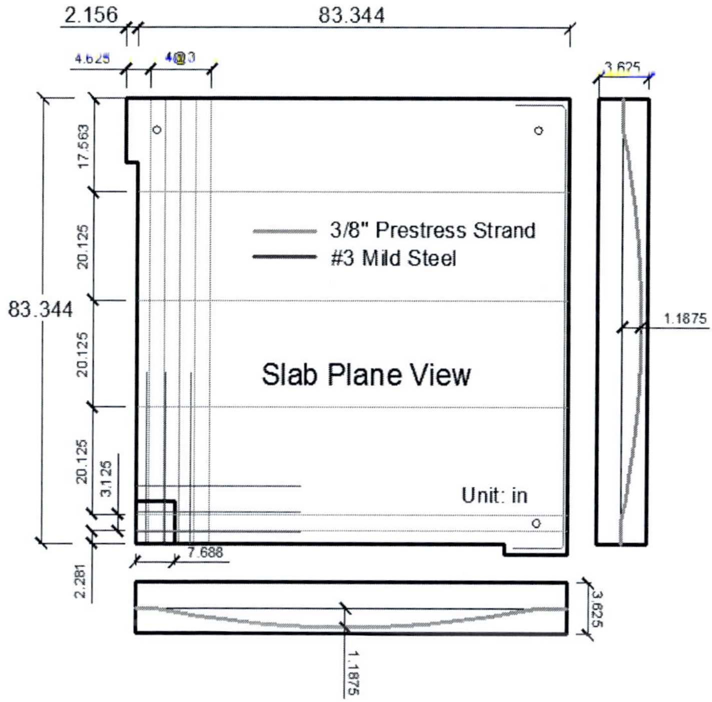

Figure 5.18 Details of reinforcement and dimensions of Specimens C1 and C2

Columns were designed to stay in the elastic range even under the largest lateral load. The half-story column below slab was pinned at its end by a universal bearing and a vertical jack was installed under the universal bearing to adjust the gravity load during testing. Another half-story column above slab was pinned at an additional universal bearing which connects to two actuators at north-south, east-west directions respectively. Slabs were pinned at three corners by vertical struts other than the location of the column to simulate the boundary condition at inflection points. A torsional restrain frame was installed parallel to the north-south direction in order to minimize the slab in-plane torsion when the lateral loading was carried out. Several lead ingots were placed at calculated positions to simulate required gravity load at initial. A schematic view and instruments setup are illustrated in Figure 5.19. A cloverleaf displacement loading pattern was applied to the column top for several drift ratio cycles in order to simulate the seismic loading. Experimental design drift ratios of 0.1%,

0.2%, 0.5%, 1.0%, 1.5%, 3.0% were used for C1 and 0.2%, 0.4%, 0.8%, 1.6%, 3.2% were applied to C2.

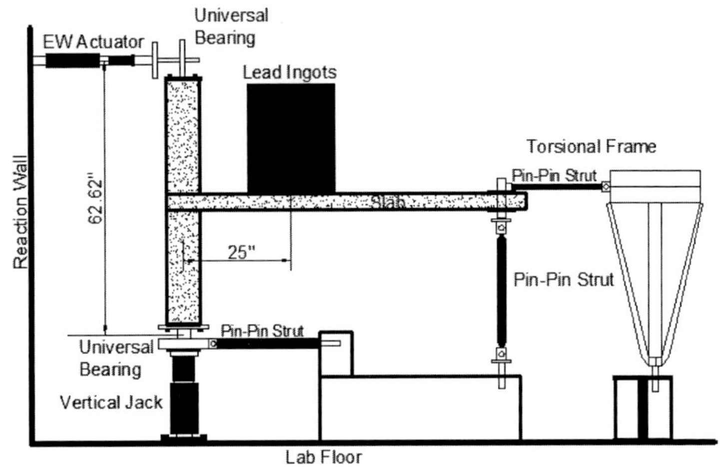

Figure 5.19 Schematic view of test setup

5.3.2 Numerical models

The modeling schemes used are the same as in the previous section, and the spring system approach is continued to be used in this section. A total of 27 springs were used for each single tendon with a total of 270 springs were used for whole slab simulation as shown in Figure 5.20.

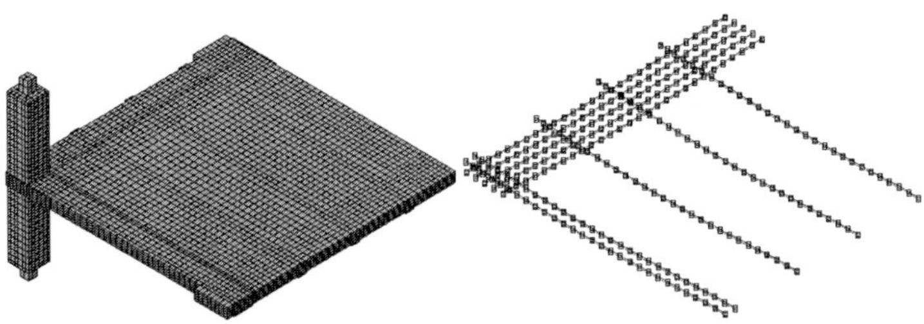

Figure 5.20 Finite element mesh and spring system of C1 and C2

Since the original experiments were conducted as cyclic loading tests, necessary measures were taken to ensure the numerical simulations are under monotonic loading conditions. Although experiments involved a cloverleaf cyclic loading pattern for both C1 and C2, the numerical study only duplicates the first two steps of certain cycles, i.e., applying a displacement loading at the column top toward the south first then changing the loading direction to west (Figure 5.21). The drift ratios chosen for numerical simulation were based on the response of lateral reaction vs. drift ratio plots in the experiment as shown in Figures 5.22 and 5.23. Possible punching shear failures occurred during the selected drift ratio loading processes. The actual drift ratios were determined from experimental response of lateral reaction vs. drift ratio plots which differs from the experimental design drift ratio a little bit. Besides Specimens C1 and C2, two additional imaginary Specimens C1-2.5 and C2-2.5 were introduced in their simulation purely based on the numerical nature. They are as exactly same as C1 and C2 except the ingot weight applied on them is 2.5 times of that on experiment. The motivation of introducing two imaginary specimens is investigating shear redistribution along the critical section under larger gravity shear. There are two sets of analyses with different drift ratios for each specimen. Therefore, four total specimens were modeled and eight simulations were carried out. The detailed information with regard to drift ratios of eight total simulations are presented in Figure 5.21.

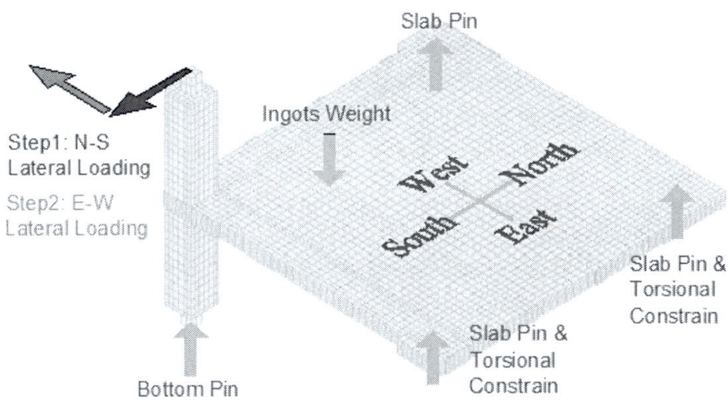

Specimens	c1a	c1a-2.5	c1b	c1b-2.5	c2a	c2a-2.5	c2b	c2b-2.5
N-S drift (%)	1.17	1.17	3.4	3.4	1.63	1.63	3.22	3.22
E-W drift (%)	1.35	1.35	3.58	3.58	1.48	1.48	3.24	3.24

Figure 5.21 Drift ratios for different numerical simulations

The effective prestress (f_{pc}) used in finite element analysis of each strand was determined from measured tendon force at the experiment. Prestress force was measured from the load cell of each tendon where load cells were installed at the slab free edge for banded tendons at opposite to the slab free edge for distributed tendons. The desired effective prestress level was achieved by several iterations of preliminary analyses via reducing the temperature field of tendon uniformly. Before lateral load analysis, three initial analysis steps were performed which were, in order, the prestressing step, self-weight step and ingot step. Prestressing was exerted by constraining only the column top and bottom because the deformation caused by prestressing was not constrained by the experimental setup. Following the prestressing step, slab corners were constrained as pin connections, as in the experiment, to be ready to sustain self-weight and ingot weight in subsequent steps. The lateral load analysis steps were performed after applying the ingot step which ensures both analytical and experimental conditions are as similar as possible.

5.3.3 Numerical results and validations

5.3.3.1 Lateral load versus drift ratio and damage pattern

Plots of lateral load versus drift ratio from C1 and C2 are compared to experimental data first. Generally, there would be some differences between experiments and analyses largely due to different loading conditions (cyclic vs. monotonic). The original tests employed cloverleaf displacement loading patterns with several cycles of different drift ratios. Each cycle contains several steps to simulate cyclic loading conditions which begin with north to south (N-S) displacement loading followed by east to west (E-W) displacement loading. These first two steps were selected in finite element analyses as monotonically increasing loadings. The discrepancy is that the experiments involved unloading for each cycle and the specimens would be damaged after one cycle test which implies initial damage for the next cycle test. The initial damage, however, was not considered in analyses. Every numerical simulation was performed without initial damage. Although these differences were present, a reasonable agreement between experiment and numerical model is expected if finite element models are well developed. Generally in this case, numerical results at the N-S loading stage are expected to comply with the backbone curves of experiments while this is not true for E-W loading stage. The simple reason is that the initial E-W lateral reaction of each drift ratio cycle is different at the E-W loading stage. The aforementioned method is not justified for the E-W loading stages.

Alternatively, a reasonable agreement of the E-W lateral reaction at the end of this loading stage is expected instead.

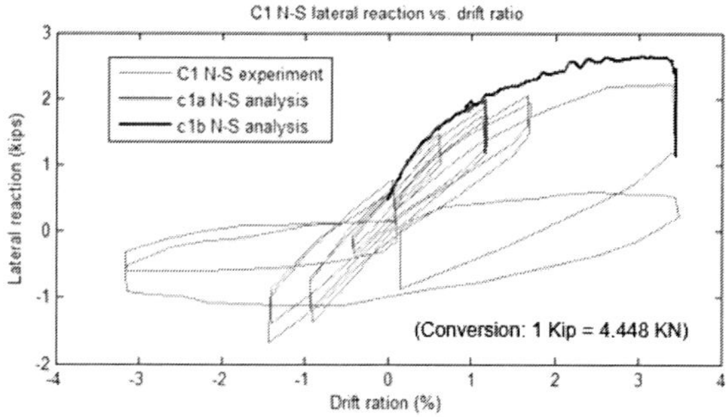

(a)

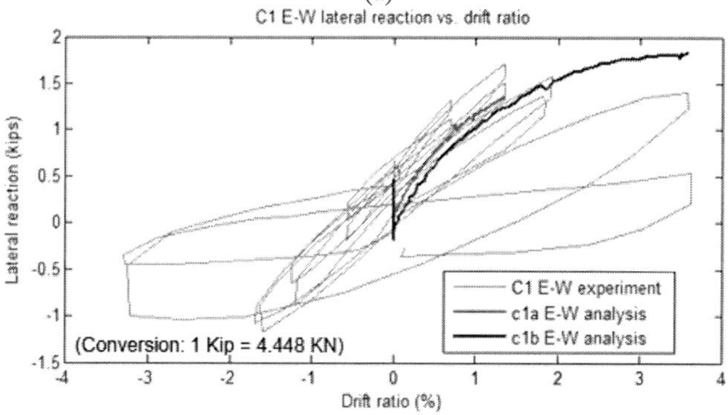

(b)

Figure 5.22 Global responses of numerical simulations related to C1

Figure 5.22 shows the N-S and E-W lateral reactions during the N-S and E-W lateral loading stages respective of c1a and c1b. The numerical results reasonably agreed with experiments at the N-S lateral loading stage. However, c1a predicted a lower lateral reaction while c1b predicted a higher lateral reaction at the end of the E-W lateral loading stage. Figure 5.23 shows the N-S and E-W lateral reactions during the N-S and E-W lateral loading stages respective of c2a and c2b. Except for c2a, which underestimated lateral reaction at the end of E-W lateral loading stage, the rest of the models have good agreement with experiments. Specimens C1 and C2 have identical geometry, are reinforced with similar prestressing and have similar material properties though C1 has initial damage due to mishandling (cracks were found on the top surface at an angle of about 45 degrees with respect to the slab free edge). In addition, C1 has been tested with 11 repetitive cycles of cloverleaf loading. C1 would be severely damaged during tests which might be another reason why discrepancies were found at the E-W lateral reaction plots. In contrast, C2 with only 5 cycles of cloverleaf loading and without initial damage had fewer differences compared to numerical results.

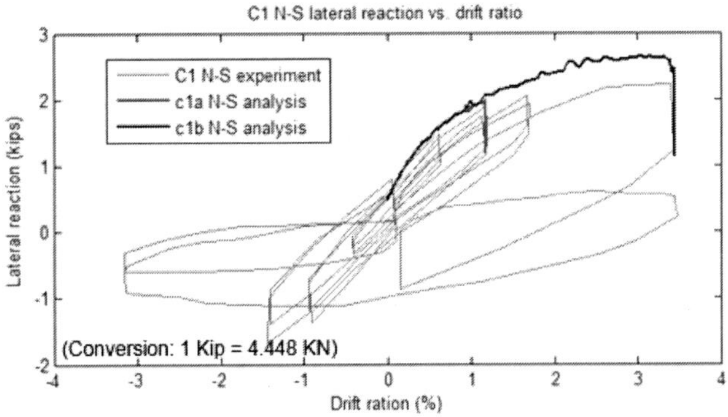

(a)

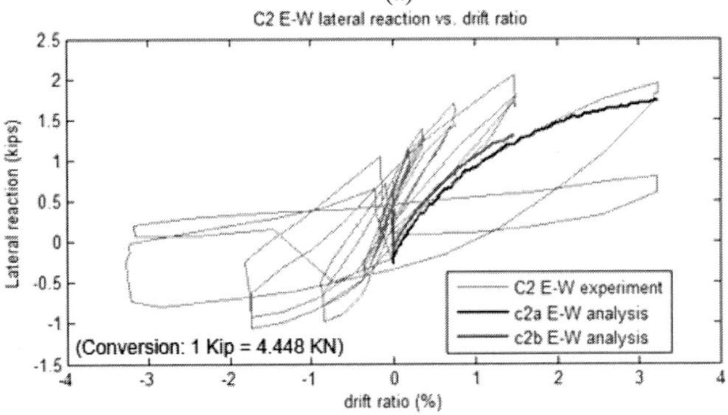

(b)

Figure 5.23 Global responses of numerical simulations related to C2

The damage patterns of simulations c1b and c2b and their experimental counterparts are shown in Figure 5.24, 5.25, 5.26, and 5.27. Because c1b and c2b are the simulations closest to the experimental conditions, only they are compared with experiments. It is noted that the numerical simulations have much less damage than their experimental counterparts. In spite of the initial damages presented in C1 (Figure 5.28), several cycles of loading and unloading might cause the excessive damages in the experiments. The complicated experimental boundary condition and its simplification in numerical simulations might have played another role in causing the observed difference. Also, the concrete constitutive model could fail to handle such a complicated tri-axial stress state of concrete. Further investigation requires a thorough study of the concrete constitutive model, complete modeling and simulation of the cyclic tests which is beyond the topic of this study.

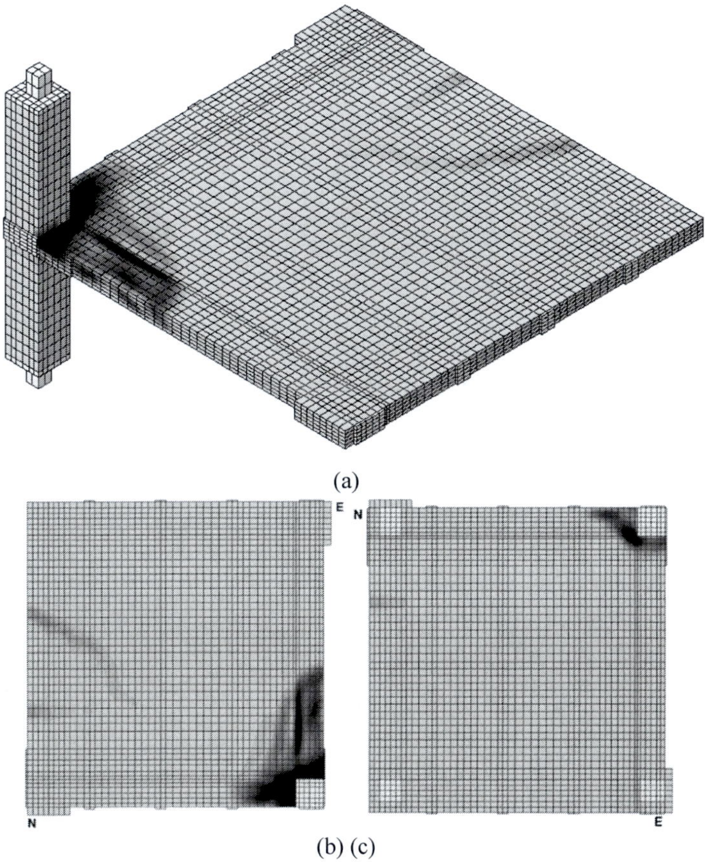

Figure 5.24 Damage pattern of simulation c1b (a) perspective view, (b) top view and (c) bottom view

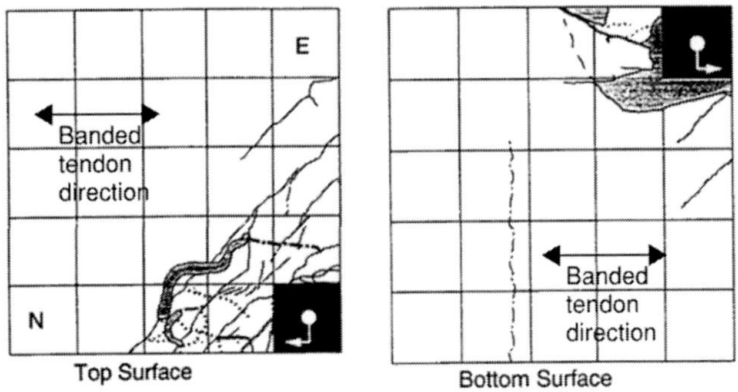

Figure 5.25 Damage pattern of C1 at failure point (Martinez-Cruzado, 1993)

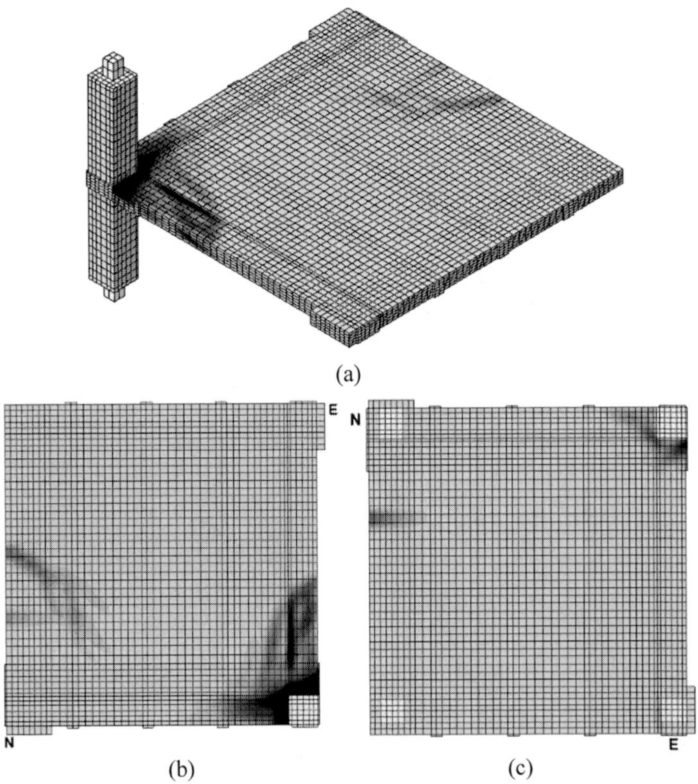

Figure 5.26 Damage pattern of simulation c2b (a) perspective view, (b) top view and (c) bottom view

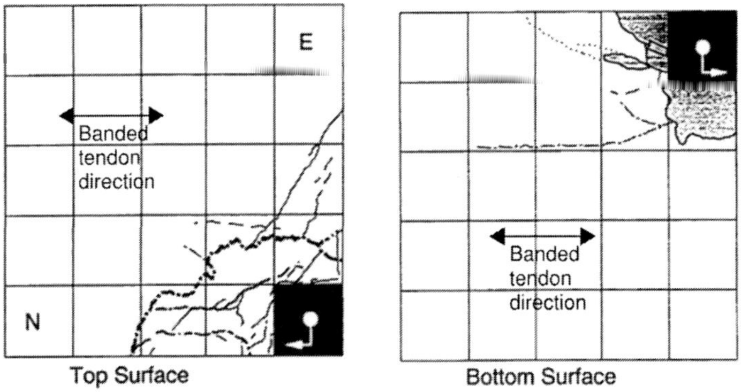

Figure 5.27 Damage pattern of C2 at failure point (Martinez-Cruzado, 1993)

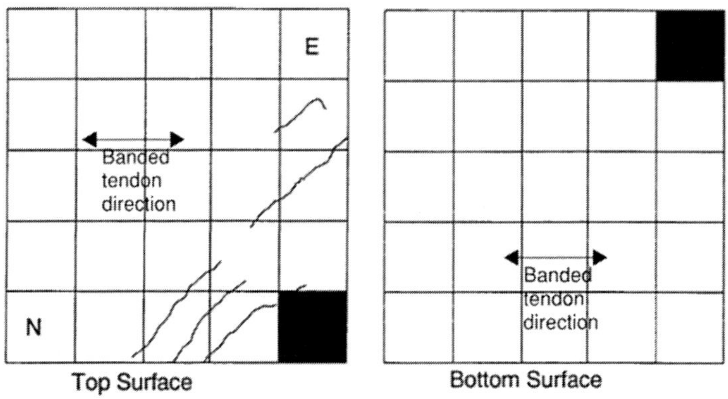

Figure 5.28 Damage pattern of C1 at initial stage (Martinez-Cruzado, 1993)

5.3.3.2 Shear distribution along critical section

ACI 318-19 describes moment and shear transfer at slab-column connections. This empiric model considers designing shear and moment transfer on a critical section located at ($d/2$) from the column face. It assumes the total shear at this section is the sum of direct shear and eccentric shear due to a fraction of unbalanced moment transfer. The shear redistribution can be predicted through this model. In order to study the shear redistribution under varying direct shear load and unbalanced moment, the shear distribution along the shear critical section was plotted based on numerical results. The vertical shear stress distribution is not addressed in the eccentric shear stress model which assumes the shear stress is independent of vertical position. In order to comply with this assumption and obtain the general pattern of the shear redistribution along the critical section, numerical shear stresses were exported from integration points of five vertical elements which are in the same location on the plan view. Shear stresses τ_{12} and τ_{23} were read from the east and north faces of the critical section respectively. The final plotted numerical shear stress is based on averaging the five layers shear stresses. Figure 5.29 shows the plan configuration of locations where average shear stresses were calculated and plotted. The East and North side of the critical section contains 8 locations to monitor the average shear.

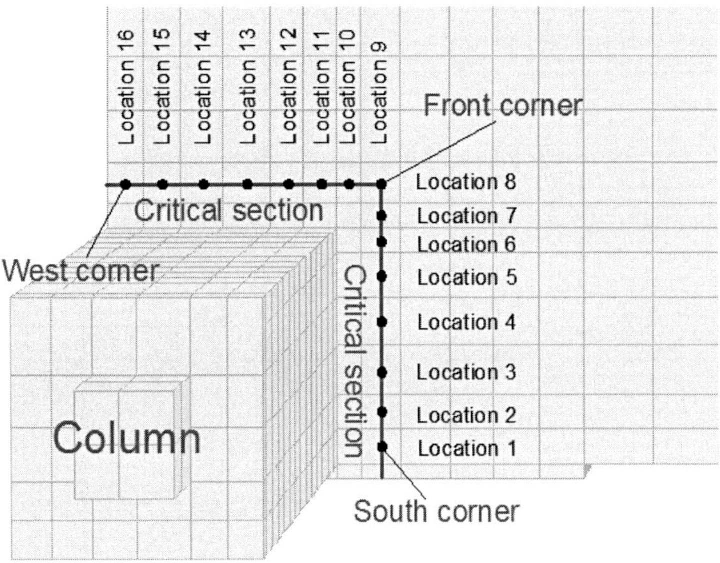

Figure 5.29 Computational points of shear stress in plan configuration

Figure 5.30 shows the shear redistribution histories of each location at the critical section of c1a analysis. The numerical shear stresses are plotted versus drift ratio which clearly presents variations of the shear stresses during lateral loadings. The shear stress histories calculated from the eccentric shear stress model at location 1 (south corner), locations 8 and 9 (front corner), and location 16 (west corner) are also included in Figure 5.30 as for comparison. It is obvious that numerical shear stress at the south corner particularly, the shear stress redistribution pattern, reasonably agrees with calculations from the eccentric shear stress model. However, a large difference of the initial shear stress is noticed at south corner. This is because the eccentric shear stress model only considers the direct shear and the unbalanced moment induced eccentric shear at initial, while for this case, initial shear stress due to prestress is considerable. Given the fact that tendons are banded in the N-S direction around column strip, the initial shear stress at the north side of the critical section due to prestress is much higher than that at the east side. As a result, the eccentric shear stress model presents a quite different value than the numerical one. The shear redistribution pattern of numerical results and the eccentric shear stress model predictions, however, are similar. Even though the numerical shear stress at the west corner does not decrease during E-W lateral loading, considering the possible influence by cracking and that the nearest two locations (location 14 and 15) both present shear stress decrease after N-S lateral loading, the shear transfer mechanism described by the eccentric shear stress model seems to be reasonable. Be advised that the shear stresses at location 8 and location 9 represent τ_{12} and τ_{23} of the front corner, respectively, which are assumed the same in the eccentric shear stress model. However, they are apparently not the same in the numerical simulation. The shear stress τ_{12} rises much more rapidly than the eccentric shear stress model prediction during the N-S lateral loading and decrease during the E-W lateral loading. Nevertheless, τ_{23} remains almost constant before the E-W lateral loading, followed by an increase during E-W lateral loading. Their patterns are quite different from each other. A possible explanation is that the gravity induced direct shear stresses are not evenly distributed; for this case, they might be concentrated at the front corner. In addition to this, the possible cracking makes the shear distribute along the critical section more randomly. Theoretically, most of them are flexural cracks around the north side of the column during the N-S lateral loading which might have most of the gravity shear transferred through the east side of the critical section. When the simulation proceeds to the E-W lateral loading, due to previous damage at the north side of the critical section, torsion induced eccentric shear stresses (shear stress caused by unbalanced moment) at this side should be smaller than the predictions of the eccentric shear stress model which assumes the elastic property of the slab-column connection. This is more evident in c1b simulation which has much larger drift ratio at the loadings of both directions.

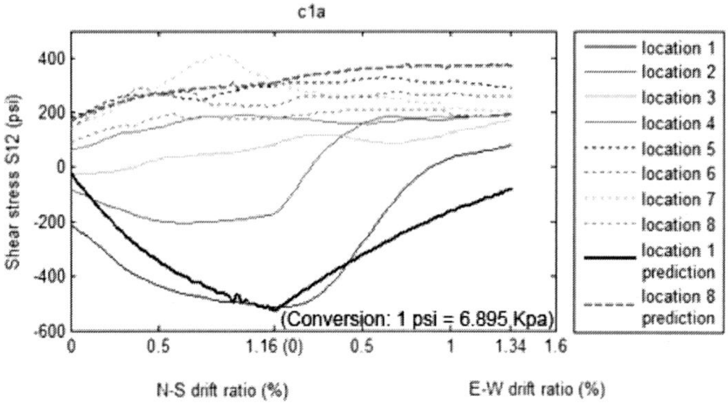

(a)

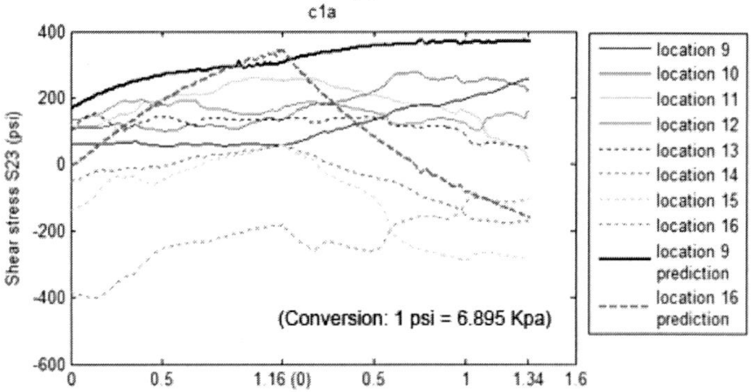

(b)

Figure 5.30 History of shear redistribution at computational points from simulation c1a

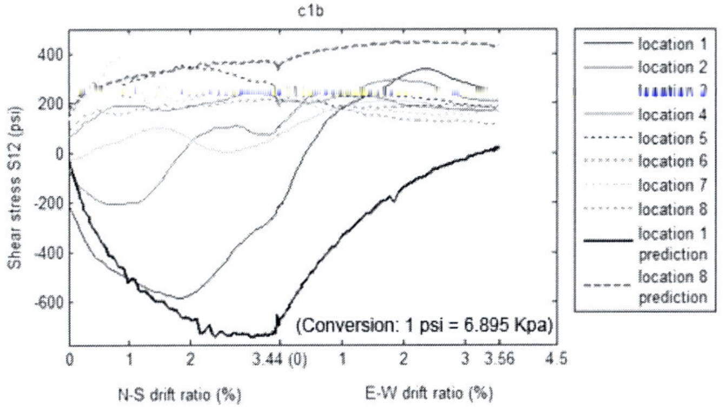

(a)

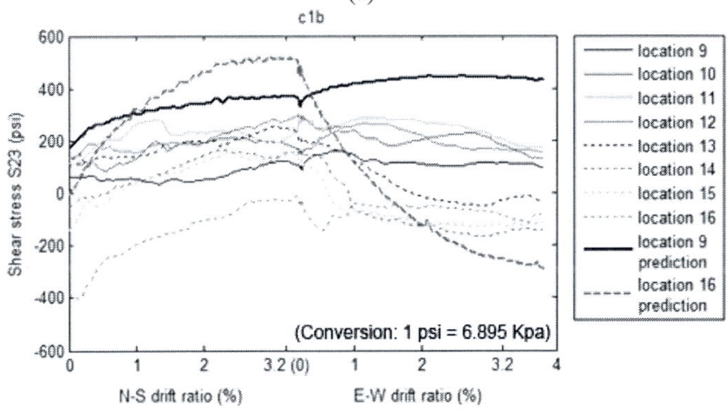

(b)

Figure 5.31 History of shear redistribution at computational points from simulation c1b

Figure 5.31 shows the same content as Figure 5.30. The stress at the front corner is much smaller than the eccentric shear stress model prediction during the whole simulation which might imply a smaller unbalanced moment transfer ratio (more discussions about unbalanced moment transfer will be presented in the following section). The numerical results at both the south and west corners present similar patterns as the c1a simulation does. In fact, numerical shear redistributions of all rest analyses (see Figures 5.32 to 5.37) behave similarly to c1a and c1b which yield conclusions like the following: 1) the eccentric shear stress model predicts the shear stress transfer and redistribution reasonably if the cracking is mildly presented; 2) the shear stress calculated followed by the eccentric shear stress model could differ too much from the real case if a large initial shear stress is presented; and 3) a reduction of the unbalanced moment transfer ratio might be considered in the calculations when the slab-column connection is severely cracked (usually at the punching shear failure stage).

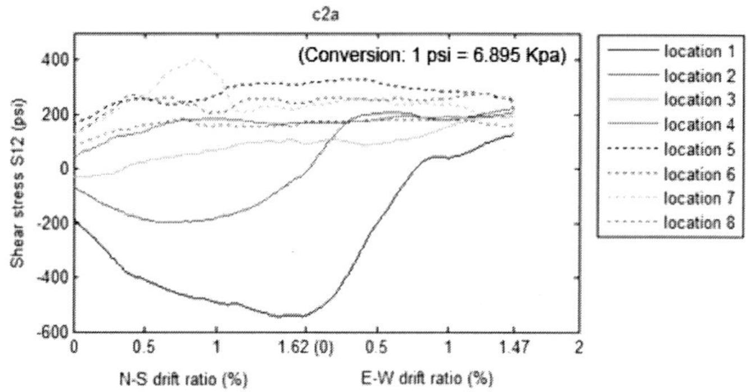

(a)

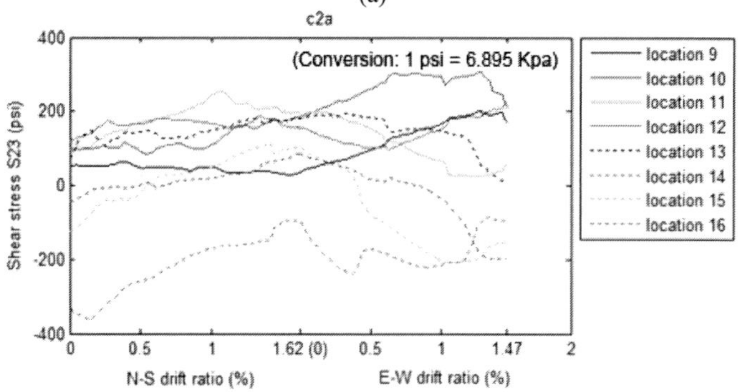

(b)

Figure 5.32 History of shear redistribution at computational points from simulation c2a

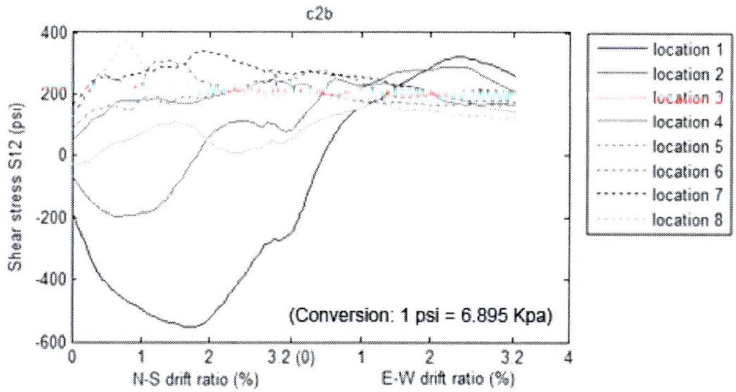

(a)

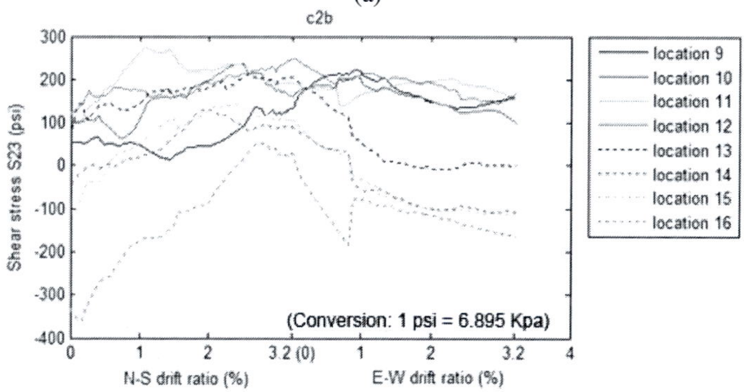

(b)

Figure 5.33 History of shear redistribution at computational points from simulation c2b

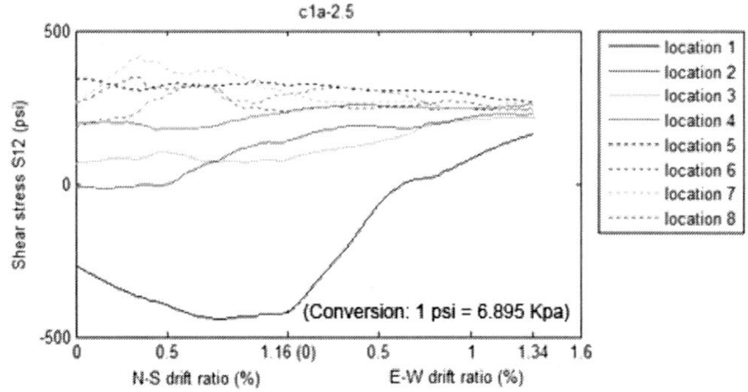

(a)

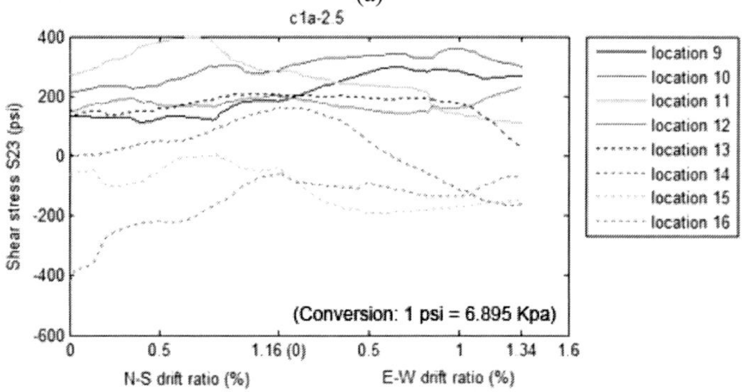

(b)

Figure 5.34 History of shear redistribution at computational points from simulation c1a-2.5

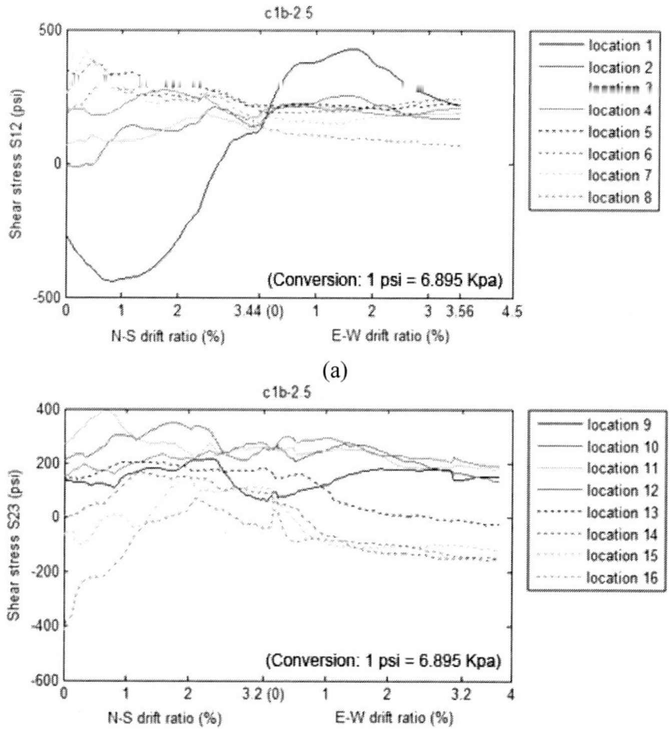

(b)

Figure 5.35 History of shear redistribution at computational points from simulation c1b-2.5

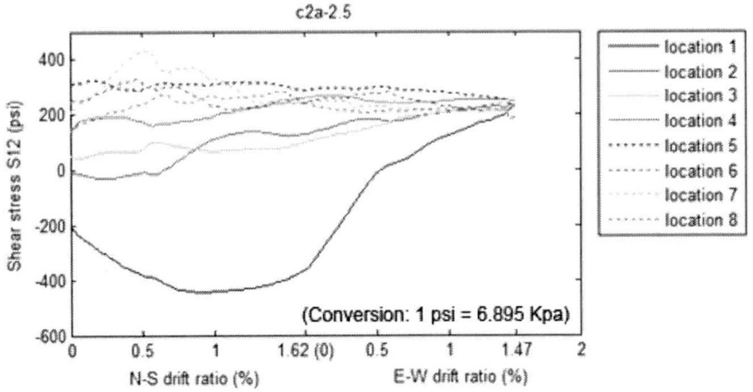

(a)

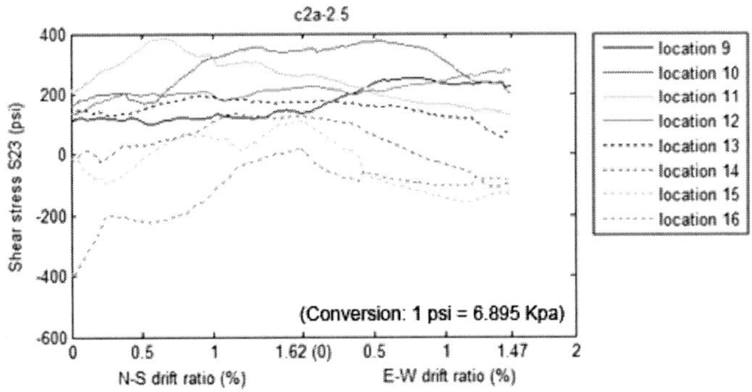

(b)

Figure 5.36 History of shear redistribution at computational points from simulation c2a-2.5

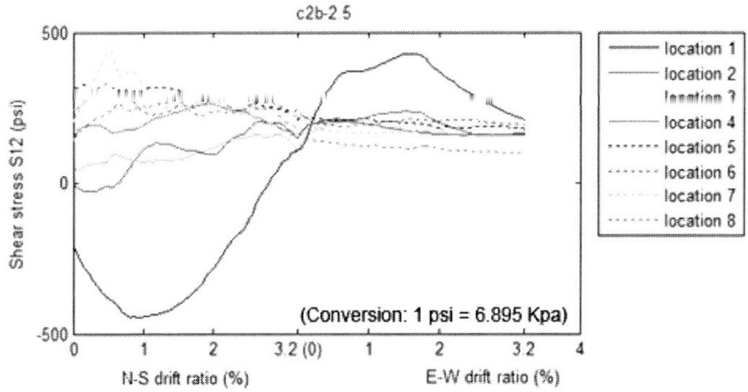

(a)

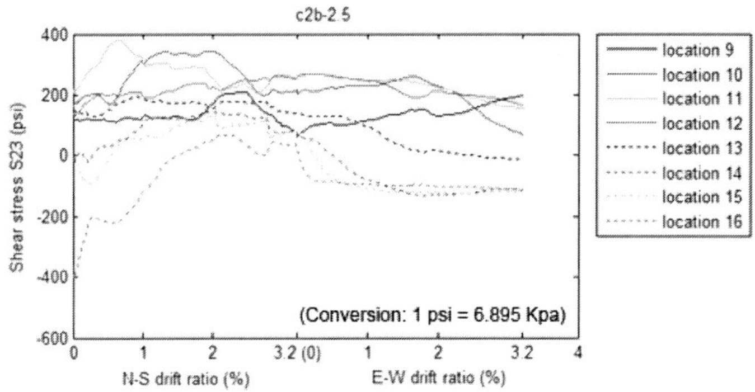

(b)

Figure 5.37 History of shear redistribution at computational points from simulation c2b-2.5

Figure 5.38 shows a three-dimensional plot of shear stress distribution along the critical section of each analysis. Only the shear stress distribution at the initial and end stages of N-S lateral loading and E-W lateral loading are shown in this figure.

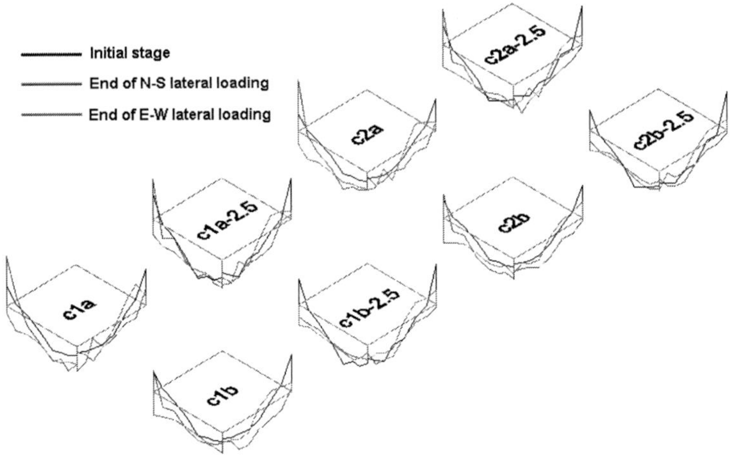

Figure 5.38 Three dimensional plots of shear redistribution of all simulations

5.3.4 Assessments of ACI 318-19 punching shear provisions

In this section, the description of the eccentric shear stress model is firstly presented, and then, the approach of calculating the design shear stress is introduced. Equations are evaluated associated with numerical simulations as well as experiments. Further analyses based on comparisons among design shear stress, numerical shear stress and ACI 318-19 permitted shear stress capacity are presented in this section which will finally yield suggestions to the currently adopted eccentric shear stress model in the code. The adjustment of unbalanced moment transfer factor γ_f is not permitted for prestressed slab-column connections according to ACI 318-19. On the other hand, the prestressing effect is not taken into account when calculating the shear capacity of such corner connections as are presented in this study. In order to access the provisions above, the numerical shear stresses obtained along the critical section which will be used to derive γ_v and assess the shear capacity in the end.

5.3.4.1 Evaluations of the eccentric shear stress model

As discussed in the previous study the current building code ACI 318-19 (2019) presents a model dealing with unbalanced moment shear transfer at slab column connections. A portion of unbalanced moment is assumed to be transferred by flexure, while the rest is transferred by the eccentric shear. The fraction of unbalanced moment transferred by flexure is given by γ_f defined by ACI 318-19, Section 8.4.2.2.2, respectively, and as already shown in Equations 5.1 and 5.2.

According to the eccentric shear stress model presented in ACI 318-19, the shear stresses at critical sections for a corner connection can be determined by the following equations:

$$v_{u(A)} = \frac{V_u}{A_c} + \frac{\gamma_x(M_{u,x} - V_u g_1)c_{AB}}{J_{cx}} + \frac{\gamma_y(M_{u,y} - V_u g_2)c_{AD}}{J_{cy}} \quad (5.6)$$

$$v_{u(D)} = \frac{V_u}{A_c} + \frac{\gamma_x(M_{u,x} - V_u g_1)c_{AB}}{J_{cx}} - \frac{\gamma_y(M_{u,y} - V_u g_2)\left(c_y + \frac{d}{2} - c_{AD}\right)}{J_{cy}} \quad (5.7)$$

$$v_{u(B)} = \frac{V_u}{A_c} - \frac{\gamma_x(M_{u,x} - V_u g_1)\left(c_x + \frac{d}{2} - c_{AB}\right)}{J_{cx}} + \frac{\gamma_y(M_{u,y} - V_u g_2)c_{AD}}{J_{cy}} \quad (5.8)$$

$$\gamma_x = 1 - \frac{1}{1 + {}^2/_3\sqrt{\frac{c_y + d/2}{c_x + d/2}}} \quad (5.9)$$

$$\gamma_y = 1 - \frac{1}{1 + {}^2/_3\sqrt{\frac{c_x + d/2}{c_y + d/2}}} \quad (5.10)$$

$$A_c = (c_x + c_y + d)d \quad (5.11)$$

where d is the effective depth of the slab; and J_{cx} and J_{cy} are polar moment of inertia of whole critical section with respect to axes of x and y respectively, and other definitions of g_1, g_2, c_{AB}, c_{CD} are shown in Figure 5.39.

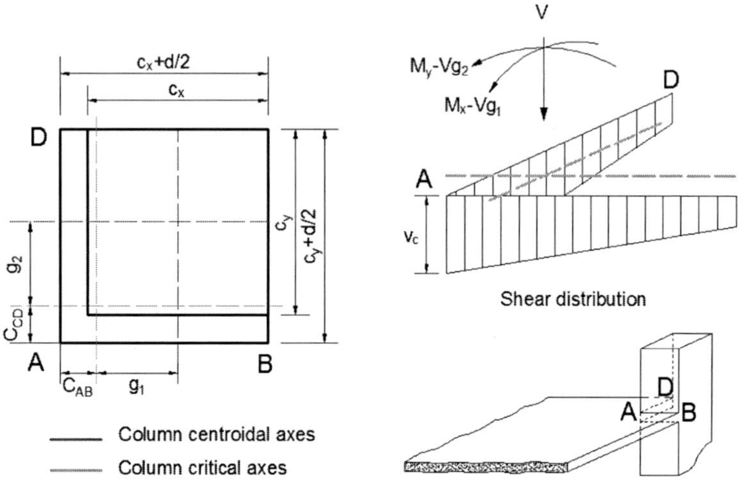

Figure 5.39 Illustration of the eccentric shear stress model

With the intention of evaluating the equations above (Equations 5.6 to 5.10), stresses calculated at the front, south and west corners of critical section (see Figure 5.29) from both experimental and numerical data are compared to stress directly obtained in numerical simulations. For simulations of c1a, c1b, c2a, c2b, two stress calculation methods were applied as: 1) stress evaluated by purely numerical data; and 2) stress evaluated mainly by experimental data along with partially numerical data. Note that some data required to perform the calculation are not accessible through experiments. For example, the experimental column reaction plots are not legible for specified cycles. Therefore they were read from numerical simulations. These factors, however, make proportionally small contribution to v_u and the differences from the experiments should be small. The evaluations mainly based on experiments thus are still valid. The shear stress v_u directly read from numerical simulation, is obtained by averaging five layers' shear stress along the vertical direction at specified locations as described in the former section. The illustration is shown in Figure 5.40.

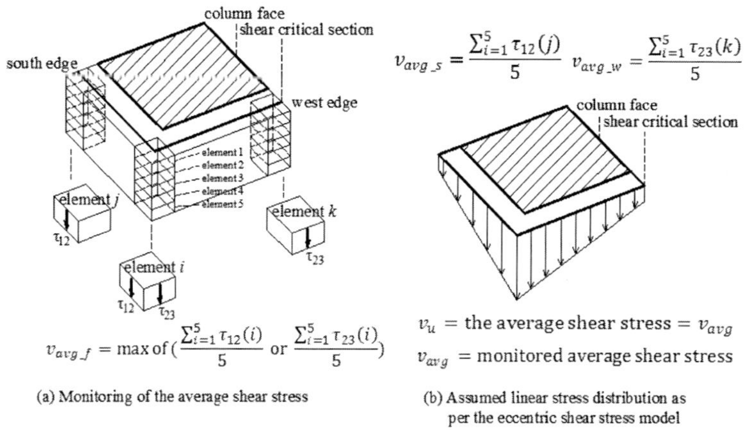

Figure 5.40 Monitoring of direct and eccentric (torsional) shear stresses associated with the eccentric shear stress model

ACI 318-19, Section 22.6.5.5 defines the punching shear strength of an interior PT slab-column connection as mentioned in Equation 5.5. Recall that Equation 5.5 is not permitted if ACI 318-19 Sections 22.6.5.5(a) and (b) are not satisfied. If the limits are not satisfied, Sections 22.6.5.2 and 22.6.5.3 shall apply which states V_c shall be the smallest of Equations 5.12, 5.13 and 5.14:

$$V_c = \left(2 + \frac{4}{\beta}\right) \lambda \sqrt{f'_c} b_0 d \qquad (5.12)$$

where β is the ratio of the long side to short side of the column, concentrated load or reaction area. λ is the modification factor reflecting the reduced mechanical properties of lightweight concrete. b_0 is the perimeter of critical section for shear in slabs. d is the effective depth of the slab.

$$V_c = \left(\frac{a_s d}{b_0} + 2\right) \lambda \sqrt{f'_c} b_0 d \qquad (5.13)$$

where a_s is 40 for interior columns, 30 for edge columns, 20 for corner columns;

$$V_c = 4\lambda\sqrt{f'_c}b_0 d \quad (5.14)$$

The specimens tested by Martinez-Cruzado (1993) do not satisfy limits (a) and (b) which implies the punching shear strength of these corner connections should be calculated using Equations 5.12 to 5.14 which is equivalent to that of an RC slab-column connection without post-tensioning. ACI 318-19 Commentary R22.6.5.4 states that the prestress is not fully effective around the perimeter of the critical section near the slab edge. Therefore the prestress is not taken into account when calculating shear capacity. Previous research (Gardner and Kallage, 1998), however, did reveal shear capacity increases even if the critical section is near prestress anchorages. The following study will compare the shear stress calculated by the eccentric shear stress model at the critical section with shear stress capacity as per ACI 318-19 Sections 22.6.5.2 and 22.6.5.3.

The focus of this numerical study is evaluating the eccentric shear stress model at the stage of punching shear failure which could produce the largest shear stress in the critical section. The ends of N-S and E-W lateral loading are considered as possible experimental punching shear failure initializations, so the calculations were made at those points. The eccentric shear stress model aims to obtain maximum possible shear stress along the critical section and the maximum stress might not necessarily happen at the ends of both N-S and E-W lateral loading during simulations. Based on the numerical data, several points other than the ends of lateral loadings are also included in the calculations where the largest shear stress occurred along the critical section. Stresses only based on numerical data are evaluated among additional points.

Table 5.2 summarizes the stress calculations at possible experimental punching shear failure points for simulations of c1a, c1b, c2a and c2b. The notations of c1a’, c1b’, c2a’ and c2b’ represents the points at the ends of N-S lateral loading for c1a, c1b, c2a and c2b simulations, respectively. The largest shear stress is marked for each case and compared with shear stress capacity v_{c1} and v_{c2} as per ACI 318-19 Sections 22.6.5.2 and 22.6.5.5. The ratio of maximum shear stress over v_{c2} indicates that the shear stress predicted by the eccentric shear stress model underestimates actual punching shear capacity (Note that the stress calculated from method 2 in Table 5.2 is mostly based on experimental data). Given the fact that the actual experimental punching shear failures of Specimens C1 and C2 did occur beyond all the points in this table which means the specimens did not really fail at these points, either the eccentric shear stress model or punching shear capacity calculation is overly conservative. As a comparison, stresses calculated by a reduced unbalanced moment transfer factor ($\gamma_v = 0.3$) are also included in the table. Even the calculations with the reduced unbalanced moment transfer factor are still larger than shear stress capacity v_{c2} in most cases. Table 5.3 summarizes the results in the same manner as Table 5.2 for simulations of c1a-2.5, c1b-2.5, c2a-2.5, and c2b-2.5.

Table 5.2 Summary of shear stresses calculated from different methods at various loading stages for simulations of c1a, c1b, c2a and c2b

possible experimental failure points		c1a			
		FEA*	FEA (max)	method1	method2
$V_{\_direct}$		NA		5766‡	5766‡
M_{x_unb}				80614‡	89089‡‡
M_{y_unb}				91888‡	115064‡‡
front	γ_v=0.4	257	258	371†	432†
south		82	-516	-81	-81
west		-104	-402	-158	-263
front*	γ_v=0.3	257	258	306†	351†
south*		82	-516	-33	-34
west*		-104	-402	-92	-170
v_{c1}		336			
v_{c2}		307			
v_u/v_{c1}		0.32			
max shear stress/v_{c2}	γ_v=0.4	front/v_{c2}		121%	141%
	γ_v=0.3	front*/v_{c2}		100%	114%

possible experimental failure points		c1a'			
		FEA*	FEA (max)	method1	method2
$V_{\_direct}$		NA		4977‡	4977‡
M_{x_unb}				131638‡	131170‡‡
M_{y_unb}				9038‡	9038‡
front	γ_v=0.4	179	193	305	304
south		-516	-516	-529†	-526†
west		-181	-402	347	346
front*	γ_v=0.3	179	193	252	251
south*		-516	-516	-373†	-371†
west*		-181	-402	284	283
v_{c1}		336			
v_{c2}		307			
v_u/v_{c1}		0.28			
max shear stress/v_{c2}	γ_v=0.4	south/v_{c2}		172%	171%
	γ_v =0.3	south*/v_{c2}		121%	121%

possible experimental failure points		c1b			
		FEA*	FEA (max)	method1	method2
$V_{\_direct}$		NA		6816‡	6816‡
M_{x_unb}				77841‡	84393‡‡
M_{y_unb}				122636‡	95903‡‡
front	γ_v=0.4	112	218	432†	394†
south		245	-583	23	-63
west		-85	-402	-293	-142
front*	γ_v=0.3	112	218	357†	328†
south*		245	-583	49	-15
west*		-85	-402	-188	-74
v_{c1}		336			
v_{c2}		307			
v_u/v_{c1}		0.38			
max shear stress/v_{c2}	γ_v=0.4	front/v_{c2}		141%	128%
	γ_v=0.3	front*/v_{c2}		116%	107%

possible experimental failure points		c1b'			
		FEA*	FEA (max)	method1	method2
$V_{\_direct}$		NA		5643[‡]	5643[‡]
M_{x_unb}				161898[‡]	146700[‡‡]
M_{y_unb}				-7[‡]	-7[‡]
front	γ_v=0.4	213	218	350	321
south		-214	-583	-685[†]	-606[†]
west		-34	-402	471	442
front*	γ_v=0.3	213	218	289	268
south*		-214	-583	-487[†]	-428[†]
west*		-34	-402	380	358
v_{c1}		336			
v_{c2}		307			
v_u/v_{c1}		0.32			
max shear stress/v_{c2}	γ_v=0.4	south/v_{c2}		223%	197%
	γ_v=0.3	south*/v_{c2}		159%	139%

<table>
<tr><td colspan="2" rowspan="2">possible experimental failure points</td><td colspan="4">c2a</td></tr>
<tr><td>FEA*</td><td>FEA (max)</td><td>method1</td><td>method2</td></tr>
<tr><td colspan="2">$V_{\_direct}$</td><td colspan="2" rowspan="3">NA</td><td>5778‡</td><td>5778‡</td></tr>
<tr><td colspan="2">M_{x_unb}</td><td>87796‡</td><td>111507‡‡</td></tr>
<tr><td colspan="2">M_{y_unb}</td><td>88593‡</td><td>119072‡‡</td></tr>
<tr><td>front</td><td rowspan="3">γ_v=0.4</td><td>172</td><td>200</td><td>379†</td><td>482†</td></tr>
<tr><td>south</td><td>132</td><td>-545</td><td>-124</td><td>-191</td></tr>
<tr><td>west</td><td>-94</td><td>-361</td><td>-127</td><td>-241</td></tr>
<tr><td>front*</td><td rowspan="3">γ_v=0.3</td><td>172</td><td>200</td><td>311†</td><td>389†</td></tr>
<tr><td>south*</td><td>132</td><td>-545</td><td>-66</td><td>-116</td></tr>
<tr><td>west*</td><td>-94</td><td>-361</td><td>-68</td><td>-153</td></tr>
<tr><td colspan="2">v_{c1}</td><td colspan="4">336</td></tr>
<tr><td colspan="2">v_{c2}</td><td colspan="4">313</td></tr>
<tr><td colspan="2">v_u/v_{c1}</td><td colspan="4">0.32</td></tr>
<tr><td rowspan="2">max shear stress/vc2</td><td>γ_v=0.4</td><td>front/v_{c2}</td><td></td><td>121%</td><td>154%</td></tr>
<tr><td>γ_v=0.3</td><td>front*/v_{c2}</td><td></td><td>99%</td><td>124%</td></tr>
</table>

possible experimental failure points		c2a'			
		FEA*	FEA (max)	method1	method2
$V_{\_direct}$		NA		4946‡	4946‡
M_{x_unb}				143436‡	148954‡‡
M_{y_unb}				369‡	369‡
front	γ_v=0.4	175	183	311	321
south		-542	-545	-608†	-637†
west		-91	-361	414	424
front*	γ_v=0.3	175	183	256	264
south*		-542	-545	-433†	-454†
west*		-91	-361	334	342
v_{c1}		336			
v_{c2}		313			
v_u/v_{c1}		0.28			
max shear stress/v_{c2}	γ_v=0.4	south/v_{c2}		194%	203%
	γ_v=0.3	south*/v_{c2}		138%	145%

<table>
<tr><td colspan="2" rowspan="2">possible experimental failure points</td><td colspan="4">c2b</td></tr>
<tr><td>FEA*</td><td>FEA (max)</td><td>method1</td><td>method2</td></tr>
<tr><td colspan="2">$V_{\_direct}$</td><td colspan="2" rowspan="3">NA</td><td>6395‡</td><td>6395‡</td></tr>
<tr><td colspan="2">M_{x_unb}</td><td>84882‡</td><td>83015‡‡</td></tr>
<tr><td colspan="2">M_{y_unb}</td><td>116584‡</td><td>129717‡‡</td></tr>
<tr><td>front</td><td rowspan="3">γ_v=0.4</td><td>164</td><td>223</td><td>431†</td><td>453†</td></tr>
<tr><td>south</td><td>258</td><td>-552</td><td>-38</td><td>-3</td></tr>
<tr><td>west</td><td>-164</td><td>-361</td><td>-261</td><td>-332</td></tr>
<tr><td>front*</td><td rowspan="3">γ_v=0.3</td><td>164</td><td>223</td><td>354†</td><td>370†</td></tr>
<tr><td>south*</td><td>258</td><td>-552</td><td>2</td><td>28</td></tr>
<tr><td>west*</td><td>-164</td><td>-361</td><td>-165</td><td>-219</td></tr>
<tr><td colspan="2">v_{c1}</td><td colspan="4">336</td></tr>
<tr><td colspan="2">v_{c2}</td><td colspan="4">313</td></tr>
<tr><td colspan="2">v_u/v_{c1}</td><td colspan="4">0.36</td></tr>
<tr><td rowspan="2">max shear stress/v_{c2}</td><td>γ_v=0.4</td><td>front/v_{c2}</td><td></td><td>138%</td><td>145%</td></tr>
<tr><td>γ_v=0.3</td><td>front*/v_{c2}</td><td></td><td>113%</td><td>118%</td></tr>
</table>

possible experimental failure points		c2b'			
		FEA*	FEA (max)	method1	method2
$V_{\_direct}$		NA		5200[‡]	5200[‡]
M_{x_unb}				165440[‡]	146387[‡‡]
M_{y_unb}				-6531[‡]	-6531[‡]
front	γ_v=0.4	214	220	341	305
south		-244	-552	-729[†]	-629[†]
west		29	-361	499	463
front*	γ_v=0.3	214	220	280	253
south*		-244	-552	-522[†]	-448[†]
west*		29	-361	399	371
v_{c1}		336			
v_{c2}		313			
v_u/v_{c1}		0.29			
max shear stress/v_{c2}	γ_v=0.4	south/v_{c2}		233%	201%
	γ_v=0.3	south*/v_{c2}		167%	143%

‡ : Value from FEA
‡‡ : Value from experiment
† : Governing stress
$v_{c1} = 3.5\sqrt{f_c'} + 0.3f_{pc}$ $v_{c2} = 4\sqrt{f_c'}$
Unit: lb – in.

Table 5.3 Summary of shear stresses calculated from different methods at various loading stages for simulations of c1a-2.5, c1b-2.5, c2a-2.5 and c2b-2.5

possible experimental failure points		c1a-2.5		
		FEA*	FEA (max)	method1
$V_{\_direct}$		NA		8732
M_{x_unb}				99635
M_{y_unb}				105246
Front	γ_v=0.4	266	328	455[†]
South		165	-443	-69
West		-74	-418	-106
front*	γ_v=0.3	266	328	382[†]
south*		165	-443	-11
west*		-74	-418	-38
v_{c1}		336		
v_{c2}		307		
v_u/v_{c1}		0.49		
max shear stress/v_{c2}	γ_v=0.4	front/v_{c2}		148%
	γ_v=0.3	front*/v_{c2}		125%

<table>
<tr><td colspan="2" rowspan="2">possible experimental failure points</td><td colspan="3">c1a-2.5'</td></tr>
<tr><td>FEA*</td><td>FEA (max)</td><td>method1</td></tr>
<tr><td colspan="2">$V_{\_direct}$</td><td colspan="2" rowspan="3">NA</td><td>8103</td></tr>
<tr><td colspan="2">M_{x_unb}</td><td>151654</td></tr>
<tr><td colspan="2">M_{y_unb}</td><td>28140</td></tr>
<tr><td>front</td><td rowspan="3">γ_v=0.4</td><td>295</td><td>328</td><td>403</td></tr>
<tr><td>south</td><td>-418</td><td>-443</td><td>-507†</td></tr>
<tr><td>west</td><td>-47</td><td>-418</td><td>376</td></tr>
<tr><td>front*</td><td rowspan="3">γ_v=0.3</td><td>295</td><td>328</td><td>340</td></tr>
<tr><td>south*</td><td>-418</td><td>-443</td><td>-342†</td></tr>
<tr><td>west*</td><td>-47</td><td>-418</td><td>320</td></tr>
<tr><td colspan="2">v_{c1}</td><td colspan="3">336</td></tr>
<tr><td colspan="2">v_{c2}</td><td colspan="3">307</td></tr>
<tr><td colspan="2">v_u/v_{c1}</td><td colspan="3">0.45</td></tr>
<tr><td rowspan="2">max shear stress/v_{c2}</td><td>γ_v=0.4</td><td>south/v_{c2}</td><td></td><td>165%</td></tr>
<tr><td>γ_v=0.3</td><td>south*/v_{c2}</td><td></td><td>111%</td></tr>
</table>

<table>
<tr><td colspan="2" rowspan="2">possible experimental failure points</td><td colspan="3">c1b-2.5</td></tr>
<tr><td>FEA*</td><td>FEA (max)</td><td>method1</td></tr>
<tr><td colspan="2">$V_{\_direct}$</td><td colspan="2" rowspan="3">NA</td><td>8805</td></tr>
<tr><td colspan="2">M_{x_unb}</td><td>47583</td></tr>
<tr><td colspan="2">M_{y_unb}</td><td>126484</td></tr>
<tr><td>front</td><td rowspan="3">γ_v=0.4</td><td>153</td><td>328</td><td>397†</td></tr>
<tr><td>south</td><td>218</td><td>-441</td><td>246</td></tr>
<tr><td>west</td><td>-151</td><td>-418</td><td>-314</td></tr>
<tr><td>front*</td><td rowspan="3">γ_v=0.3</td><td>153</td><td>328</td><td>339†</td></tr>
<tr><td>south*</td><td>218</td><td>-441</td><td>226</td></tr>
<tr><td>west*</td><td>-151</td><td>-418</td><td>-194</td></tr>
<tr><td colspan="2">v_{c1}</td><td colspan="3">336</td></tr>
<tr><td colspan="2">v_{c2}</td><td colspan="3">307</td></tr>
<tr><td colspan="2">v_u/v_{c1}</td><td colspan="3">0.49</td></tr>
<tr><td rowspan="2">max shear stress/v_{c2}</td><td>γ_v=0.4</td><td>front/v_{c2}</td><td></td><td>129%</td></tr>
<tr><td>γ_v=0.3</td><td>front*/v_{c2}</td><td></td><td>110%</td></tr>
</table>

<table>
<tr><td colspan="2" rowspan="2">possible experimental failure points</td><td colspan="3">c1b-2.5'</td></tr>
<tr><td>FEA*</td><td>FEA (max)</td><td>method1</td></tr>
<tr><td colspan="2">$V_{\_direct}$</td><td colspan="2" rowspan="3">NA</td><td>8014</td></tr>
<tr><td colspan="2">M_{x_unb}</td><td>121976</td></tr>
<tr><td colspan="2">M_{y_unb}</td><td>43242</td></tr>
<tr><td>front</td><td rowspan="3">γ_v=0.4</td><td>144</td><td>328</td><td>374†</td></tr>
<tr><td>south</td><td>123</td><td>-441</td><td>-279</td></tr>
<tr><td>west</td><td>24</td><td>-418</td><td>248</td></tr>
<tr><td>front*</td><td rowspan="3">γ_v=0.3</td><td>144</td><td>328</td><td>318†</td></tr>
<tr><td>south*</td><td>123</td><td>-441</td><td>-171</td></tr>
<tr><td>west*</td><td>24</td><td>-418</td><td>224</td></tr>
<tr><td colspan="2">v_{c1}</td><td colspan="3">336</td></tr>
<tr><td colspan="2">v_{c2}</td><td colspan="3">307</td></tr>
<tr><td colspan="2">v_u/v_{c1}</td><td colspan="3">0.45</td></tr>
<tr><td rowspan="2">max shear stress/v_{c2}</td><td>γ_v=0.4</td><td>front/v_{c2}</td><td></td><td>122%</td></tr>
<tr><td>γ_v=0.3</td><td>front*/v_{c2}</td><td></td><td>104%</td></tr>
</table>

possible experimental failure points		c2a-2.5		
		FEA*	FEA (max)	method1
$V_{\_direct}$		NA		8470
M_{x_unb}				103285
M_{y_unb}				99775
front	γ_v=0.4	233	283	450[†]
south		228	-446	-106
west		-86	-401	-78
front*	γ_v=0.3	233	283	377[†]
south*		228	-446	-40
west*		-86	-401	-19
v_{c1}		336		
v_{c2}		313		
v_u/v_{c1}		0.48		
max shear stress/v_{c2}	γ_v=0.4	front/v_{c2}		144%
	γ_v=0.3	front*/v_{c2}		120%

possible experimental failure points		c2a-2.5'		
		FEA*	FEA (max)	method1
$V_{\_direct}$		NA		7765
M_{x_unb}				160048
M_{y_unb}				14749
Front	γ_v=0.4	239	283	391
South		-366	-446	-586[†]
West		19	-401	452
front*	γ_v=0.3	239	283	330
south*		-366	-446	-403[†]
west*		19	-401	375
v_{c1}		336		
v_{c2}		313		
v_u/v_{c1}		0.44		
max shear stress/v_{c2}	γ_v=0.4	south/v_{c2}		187%
	γ_v=0.3	south*/v_{c2}		129%

possible experimental failure points		c2b-2.5		
		FEA*	FEA (max)	method1
$V_{\_direct}$		NA		8426
M_{x_unb}				58752
M_{y_unb}				122959
Front	γ_v=0.4	196	276	409†
South		210	-446	170
West		-116	-401	-285
front*	γ_v=0.3	196	276	346†
south*		210	-446	167
west*		-116	-401	-174
v_{c1}		336		
v_{c2}		313		
v_u/v_{c1}		0.47		
max shear stress/v_{c2}	γ_v=0.4	front/v_{c2}		130%
	γ_v=0.3	front*/v_{c2}		111%

possible experimental failure points		c2b-2.5'		
		FEA*	FEA (max)	method1
$V_{\_direct}$		NA		7762
M_{x_unb}				134594
M_{y_unb}				37582
Front	γ_v=0.4	173	276	386
South		108	-446	-409†
West		23	-401	284
front*	γ_v=0.3	173	276	326†
south*		108	-446	-270
west*		23	-401	250
v_{c1}		336		
v_{c2}		313		
v_u/v_{c1}		0.44		
max shear stress/v_{c2}	γ_v=0.4	south/v_{c2}		131%
	γ_v=0.3	front*/v_{c2}		104%

† : Governing stress
$v_{c1} = 3.5\sqrt{f_c'} + 0.3f_{pc}$ $v_{c2} = 4\sqrt{f_c'}$
Unit: lb – in.

It is expected that with presence of a higher gravity load, stress at the front corner increases while stresses at the south and west corners decrease as a result of increase of direct shear stress. Even under higher gravity shear stress, the eccentric shear stress model and the punching shear capacity are still conservative. Also stresses calculated from reduced γ_v are included in the table for comparison which indicates that a smaller γ_v (= 0.3) might be more appropriate for square corner connections. The 'FEA*' and 'FEA(max)' represent the current and maximum shear stress readings, respectively, during the analysis. The larger discrepancy between 'FEA*' and 'FEA(max)' indicates the fact that the maximum numerical shear stress may not occur right at the ending of lateral loadings. The maximum shear stress actually occurred during the N-S lateral loading for all c1, c1-2.5, c2 and c2-2.5 simulations. Figures 5.41(a) and (b) show the maximum stress always occurs at south corner during the N-S lateral loading.

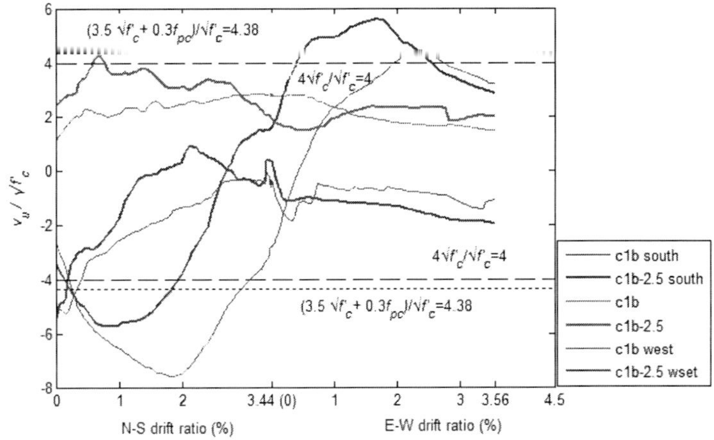

(a)

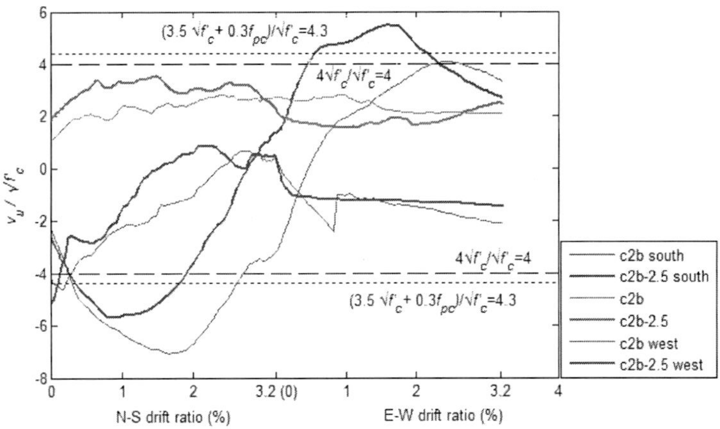

(b)

Figure 5.41 Normalized numerical shear stress vs. drift ratio at three critical points (from numerical simulations of c1b, c1b-2.5, c2b and c2b-2.5)

It might be more proper to evaluate the eccentric shear stress model when the maximum shear stress appears in the critical section. Table 5.4 shows the evaluations made at those points, where the 'FEA*' and 'FEA(max)' are the same. Nevertheless, considering that the large initial shear stresses caused by prestress at the south and west corners is not considered in the eccentric shear stress model, the predicted shear stresses by the eccentric shear stress model would be much higher than the numerical shear stresses. Therefore, the results still yield the same conclusion that either the eccentric shear stress model or the shear stress capacity as per ACI 318-19 Section 22.6.5.2 is conservative. The next section will discuss more about unbalanced moment transfer ratio (γ_v).

Table 5.4 Summary of shear stresses calculated from different methods at various loading stages for c1b, c1b-2.5, c2b and c2b-2.5

possible experimental failure points		c1b		
		FEA*	FEA (max)	method1
$V_{\_direct}$		NA		5069
M_{x_unb}				137729
M_{y_unb}				6560
Front	γ_v=0.4	180	198	347
South		-583	-583	-665†
West		-113	-402	449
front*	γ_v=0.3	180	198	286
south*		-583	-583	-473†
west*		-113	-402	362
v_{c1}		336		
v_{c2}		307		
v_u/v_{c1}		0.49		
max shear stress/v_{c2}	γ_v=0.4	south/v_{c2}		217%

<table>
<tr><td colspan="2" rowspan="2">possible experimental failure points
$V_{\_direct}$</td><td colspan="3">c1b-2.5</td></tr>
<tr><td>FEA*</td><td>FEA (max)</td><td>method1
7999</td></tr>
<tr><td colspan="2">M_{x_unb}</td><td colspan="2" rowspan="2">NA</td><td>145276</td></tr>
<tr><td colspan="2">M_{y_unb}</td><td>29269</td></tr>
<tr><td>Front</td><td rowspan="3">γ_v=0.4</td><td>271</td><td>328</td><td>392</td></tr>
<tr><td>South</td><td>-441</td><td>-441</td><td>-453†</td></tr>
<tr><td>West</td><td>-167</td><td>-418</td><td>329</td></tr>
<tr><td>front*</td><td rowspan="3">γ_v=0.3</td><td>271</td><td>328</td><td>331†</td></tr>
<tr><td>south*</td><td>-441</td><td>-441</td><td>-302</td></tr>
<tr><td>west*</td><td>-167</td><td>-418</td><td>284</td></tr>
<tr><td colspan="2">v_{c1}</td><td colspan="3">336</td></tr>
<tr><td colspan="2">v_{c2}</td><td colspan="3">307</td></tr>
<tr><td colspan="2">v_u/v_{c1}</td><td colspan="3">0.45</td></tr>
<tr><td>max shear stress/v_{c2}</td><td>γ_v=0.4</td><td>south/v_{c2}</td><td></td><td>147%</td></tr>
</table>

possible experimental failure points		c2b		
		FEA*	FEA (max)	method1
$V_{\_direct}$		NA		4434
M_{x_unb}				115002
M_{y_unb}				11431
Front	γ_v=0.4	186	196	315
South		-552	-552	-631†
West		-102	-361	439
front*	γ_v=0.3	186	196	260
south*		-552	-552	-449†
west*		-102	-361	353
v_{c1}		336		
v_{c2}		307		
v_u/v_{c1}		0.5		
max shear stress/v_{c2}	γ_v=0.4	south/v_{c2}		206%

possible experimental failure points		c2b-2.5		
		FEA*	FEA (max)	method1
$V_{\_direct}$		NA		7404
M_{x_unb}				137962
M_{y_unb}				27441
Front	γ_v=0.4	258	266	374
South		-446	-446	-468†
West		-159	-401	344
front*	γ_v=0.3	258	266	316†
south*		-446	-446	-316
west*		-159	-401	293
v_{c1}		336		
v_{c2}		307		
v_u/v_{c1}		0.45		
max shear stress/v_{c2}	γ_v=0.4	south/v_{c2}		153%

† : Governing stress

$v_{c1} = 3.5\sqrt{f_c'} + 0.3f_{pc}$ $v_{c2} = 4\sqrt{f_c'}$

Unit: lb – in.

5.3.4.2 Assessment of the unbalanced moment transfer factor and the punching shear capacity

The assessment of γ_v is accomplished by substituting numerical shear stress v_u into Equation 5.6. This assessment is purely based on numerical results. Because the initial shear stresses are very large at the south corner and west corner due to prestress, the initial shear stress of the west corner particularly differs from the eccentric shear stress model prediction so much, directly obtained γ_v is not reasonable. For that reason, only the front corner, which has relatively small initial shear stress due to prestress, is evaluated in this section. Figure 5.42 shows the derived γ_v from the method described above. All γ_v values show a similar pattern (see Figure 5.42). They are always smaller than 0.3 and tend to decrease as with drift ratios increase. This is consistent with the conclusion drawn from previous sections which clearly indicates that the γ_v is smaller than the eccentric shear stress model value when substantial damage is present. This is reasonable because the connection region might not be rigid elastic any more as the cracks progress. Subsequently, torsional resistance is reduced which leads to the reduction of unbalanced moment transferring by torsion. Although the γ_v is not obtained at the south and west corner, the stress calculation at previous sections implies a reduction of γ_v or increase of punching shear capacity. A conclusion can be drawn together with the analysis of this section that a reduction of γ_v to 0.3 is suggested. Even though analyses based on both experimental and numerical data lead to the same conclusion as this study, more investigations of PT slab-column corner connections are of great need for further understanding of the punching shear failure mechanism.

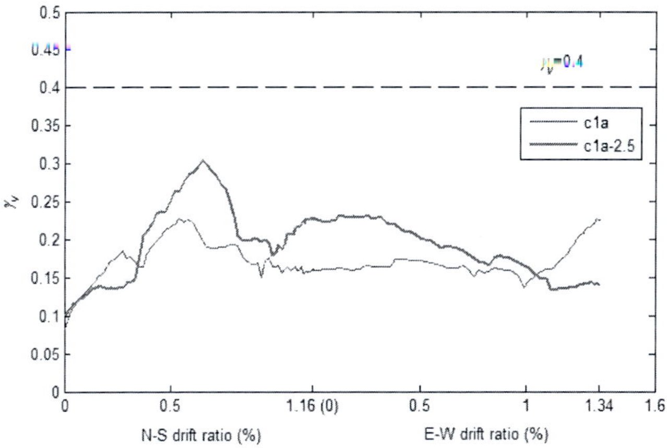
γ_v=0.4
c1a
c1a-2.5
γ_v
N-S drift ratio (%)
E-W drift ratio (%)
1.16 (0)
1.34

(a)

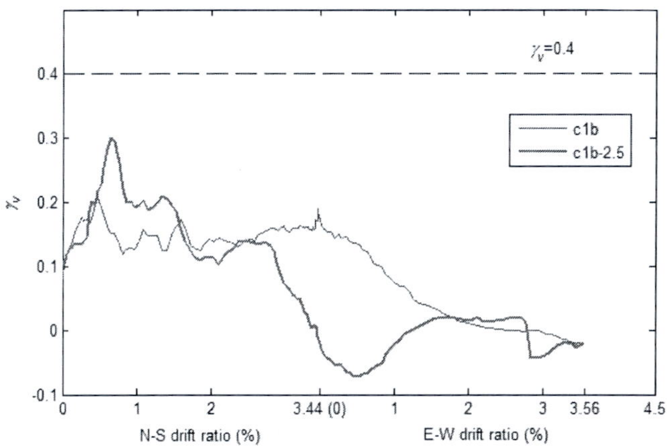
γ_v=0.4
c1b
c1b-2.5
γ_v
N-S drift ratio (%)
E-W drift ratio (%)
3.44 (0)
3.56

(b)

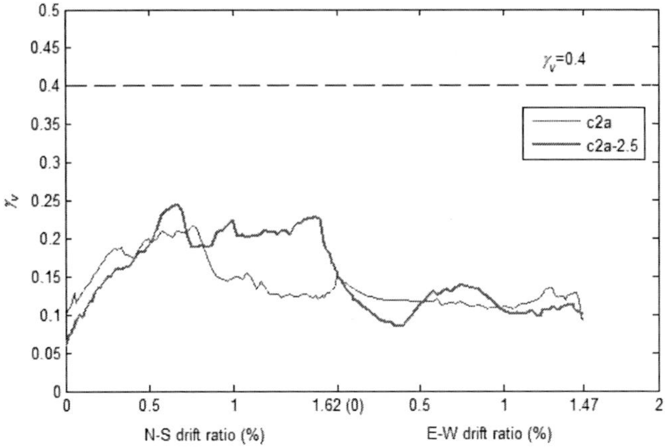

(c)

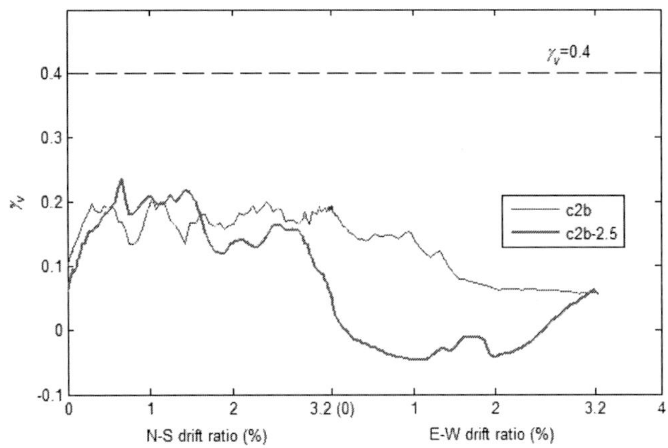

(d)

Figure 5.42 Unbalanced moment transfer ratios derived from numerical data at the front corner of each simulation

5.3.5 Conclusion

The previously proposed finite element modeling schemes dealing with unbonded PT structures have been utilized in this study to investigate ACI 318-19 punching shear provision applying to PT slab-column corner connections. Two corner specimens tested by Martinez-Cruzado were modeled. Reasonable agreements are achieved between numerical simulations and experiments while some discrepancies of Specimen C1 were expected due to initial damages. Based on the validated finite element model and the corresponding numerical simulations, the following conclusions can be drawn:

(1) The initial shear stress due to prestress should be considered when employing the eccentric shear stress model to calculate shear stress along the critical section particularly in the direction where prestressing tendons are banded.

(2) The shear stress transfer and redistribution are reasonably agreed with the eccentric shear stress model, especially in the elastic stage or with mild cracks formed in the connection. Otherwise interaction between moment and shear transfer is reduced when the connection is severely damaged.

(3) Either the eccentric shear stress model or the shear stress capacity as per ACI 318-19 Sections 8.4.2.2.2 and 8.4.4.2 are too conservative at possible failure stages. Relief of the unbalanced moment transfer factor (γ_v) and including prestress influence in the shear stress capacity at PT slab-column corner connections are suggested.

(4) Based on the eccentric shear stress model, unbalanced moment transfer factor (γ_v) can be derived from numerical data. The γ_v vs. drift ratio plot was only made at the front corner due to its relatively small initial shear stress caused by prestress. All γ_v vs. drift ratio plots yield the same conclusion that γ_v never exceeded 0.3 in these analyses and decreased as cracks proceeded. A reduction of γ_v (for example, $\gamma_v = 0.3$) for PT slab-column corner connections could be permitted. However, more investigations on different PT slab-column corner connections are desired to verify these recommendations on a more general basis.

5.4 Numerical Simulations of a Two-Way Bonded PT Interior Slab-Column Connection

5.4.1 Description of specimen

A two-way grout-bonded PT interior slab-column connection was tested by Prawatwong et al. (2007). The slab thickness was 4.7 in. The slab was extended to mid-span from the interior column and the column was extended above and below the slab to mid-height of one story, assuming that inflection points would be located at slab mid-span and column mid-height. The test slab had a footprint of 224.4 x 224.4 in. The rectangular column dimension was 19.7 x 9.8 in. with a long side aligned with the lateral load direction. The slab was supported along two slab edges in the direction transverse to the lateral load direction, where five pin connections were used to simulate the boundary condition of inflection points. They were located 94.5 in. away from the column center on each side. The column bottom end was set on a hinge connection 35.4 in. away from the center of the slab. The column top was connected to an actuator at a distance of 35.4 in. from the slab center. Grade 270 seven-wire strands with a straight tendon profile were used for both directions as shown in Figure 5.43(a). The tendons in the direction transverse to lateral loading were located at the slab mid-depth, whereas the tendons in the direction of lateral loading were placed just above the transverse tendons. A total of eight tendons were provided in both directions, banded with spacing of 13.8 in. in the loading direction and distributed in the direction perpendicular to the loading with spacing of 27.6 in. No tendons were placed passing through the column. Approximately 80% of the ultimate strength was prestressed for each tendon, which was then grouted.

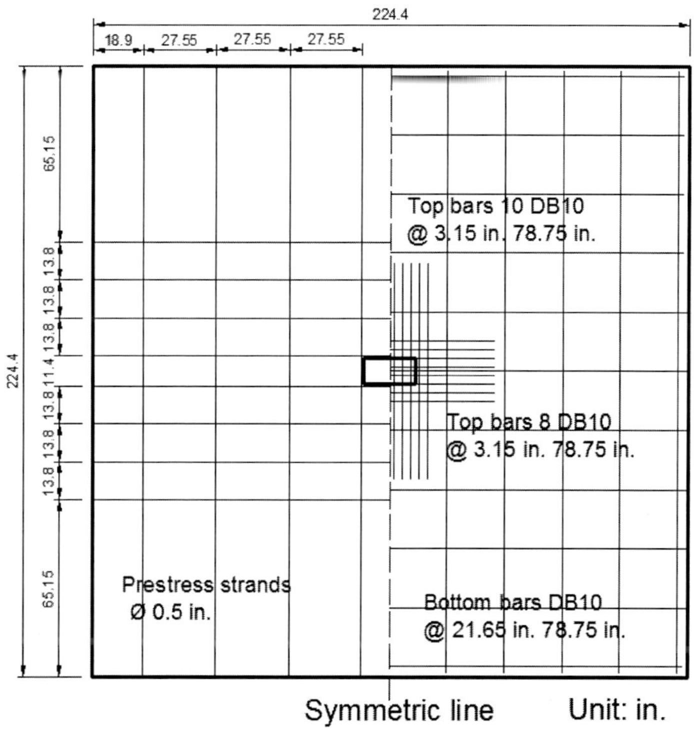

(a)

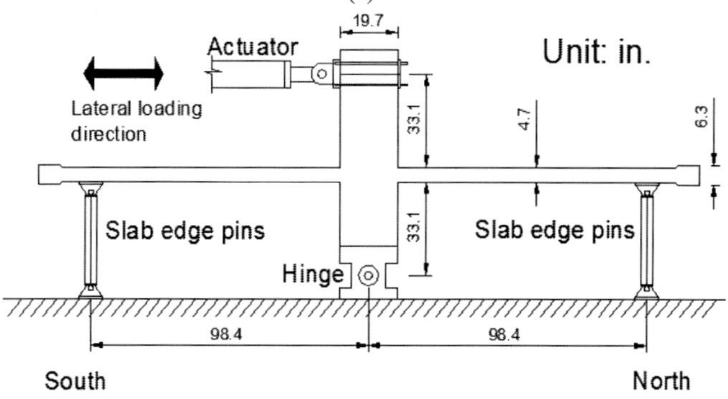

(b)

Figure 5.43 Grout-bonded PT slab-column connection; experiment conducted by Prawatwong et al. (2007)

Bonded mild steel with a diameter of 0.39 in. and length of 78.7 in. also was provided at the top and bottom at the slab-column connection region. Eight and ten top bars were placed in the loading and transverse directions, respectively, at spacing of 3.15 in. and within the width of $c_2 + 3h$, where c_2 is the column dimension of 9.8 in. in the transverse direction and h is the slab thickness. The bottom mat of the same mild steel was placed throughout the whole slab at a spacing of 21.65 in. with only one continuous bottom bar placed within the column in each direction. The layouts of post-tensioning tendons and non-prestressed reinforcement are shown in Figure 5.43(a). Additional dead loads were applied to the slab using sand bags to obtain a desired gravity shear ratio of about 0.3 at the connection before lateral testing. The torsional deformation was restrained by a torsional restraining system. A typical displacement-controlled cyclic loading test was carried out with monotonically increasing drift levels of 0.25%, 0.5%, 0.75%, 1%, 1.25%, 1.5% and 2%. Two complete cyclic displacement loops were made for each drift ratio. Figure 5.43(b) shows the test setup, with more details available elsewhere (Pimanmas et al., 2004; Warnitchai et al., 2004; Prawatwong et al., 2007).

5.4.2 Numerical model

The contact formulation was employed to model the bonded PT system which cannot be simulated through the spring system method. As mentioned in the previous chapter, the prestressing stage was simulated under frictionless contact which yields a perfect unbonded condition. After the prestressing, the tangential constitutive model of contact was switched to 'tough' where perfect bonding is achieved between the tendon and concrete. Typical dimension of elements in the modeling is 1 in. and five layers were arranged in the slab thickness direction as shown in Figure 5.44. A similar analysis approach used in the previous study towards to corner connections was applied here to treat the cyclic loading condition. Therefore, the column top was applied only by a monotonically increased displacement until the target drift ratio of 2% was reached in the simulations. For the purpose of validation of modeling schemes, the numerical global response of the specimen was compared to the backbone curve of the hysteresis loops obtained from the experiment. Furthermore, local behavior such as strains in mild steel bars and tendons were also compared to experimental data. In order to study the bonding influence of prestressing tendons, an identical imaginary specimen but with unbonded PT tendons was simulated and compared with the bonded counterpart in terms of flexural and punching shear behaviors.

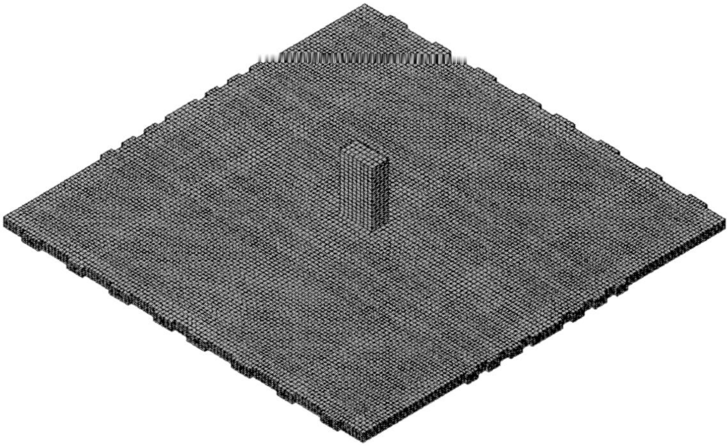

(a)

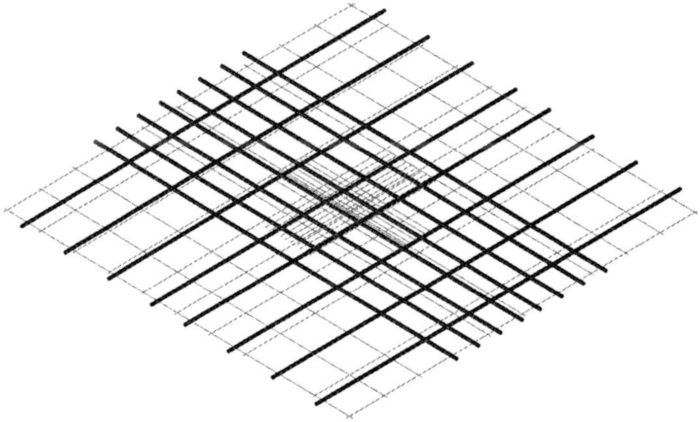

(b)

Figure 5.44 Finite element mesh and reinforcement modeling for the slab

5.4.3 Numerical results and validations

The global load-displacement responses from the numerical simulations are compared with the backbone curve of the hysteresis loops. Figure 5.45 shows a good agreement between the modeling and testing, including the lateral stiffness and strength. A slight underestimation of the post-yield stiffness is the only aberration observed. The monotonic responses from the simulations of grout-bonded PT and unbonded PT connections are almost the same. Note that the only difference for the additional simulation was the change in bonding condition of prestressing tendons.

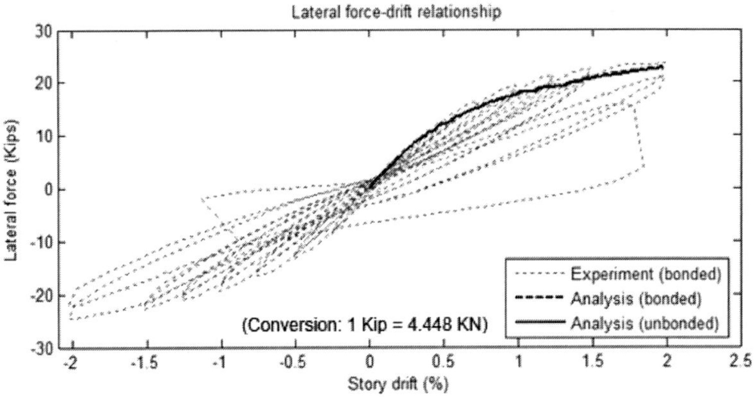

Figure 5.45 Simulated backbone curve of a two-way bonded PT interior slab-column connection; experiment conducted by Prawatwong et al. (2007)

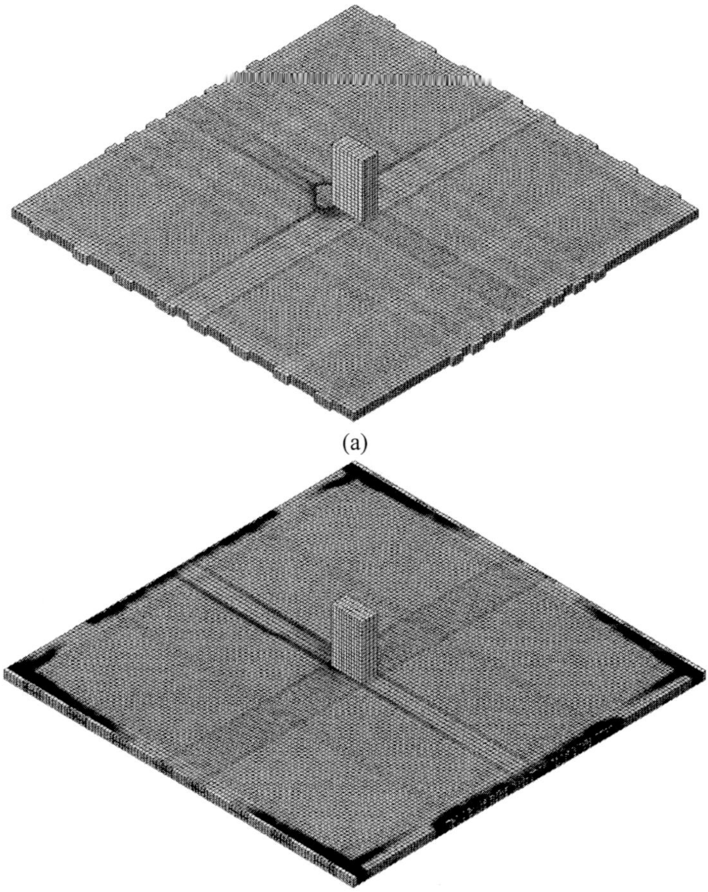

(a)

(b)

Figure 5.46 Crack pattern from perspective view; (a) bonded specimen and (b) unbonded specimen

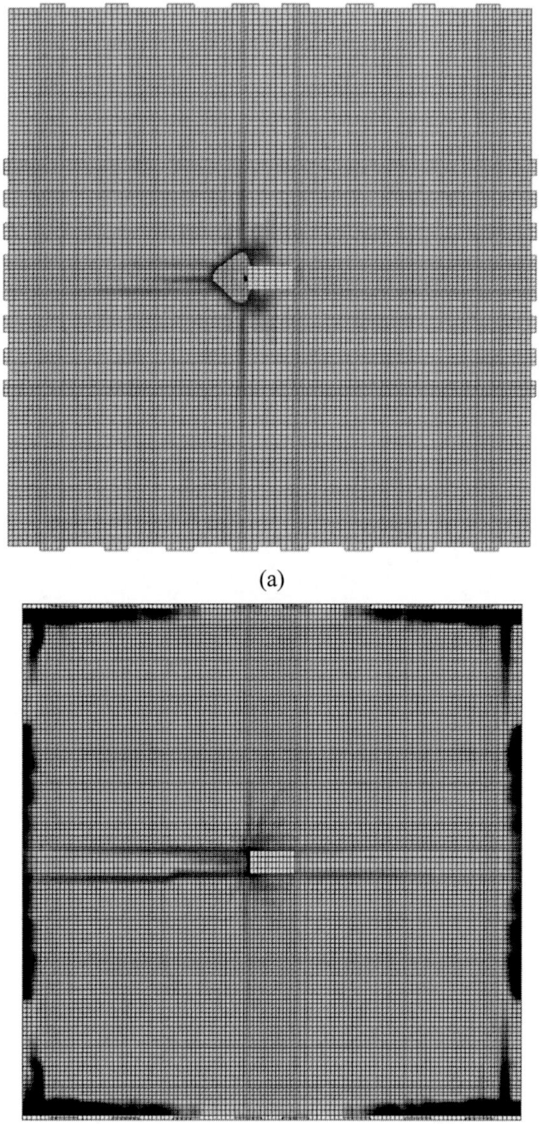

(a)

(b)

Figure 5.47 Crack pattern from top view; (a) bonded specimen and (b) unbonded specimen

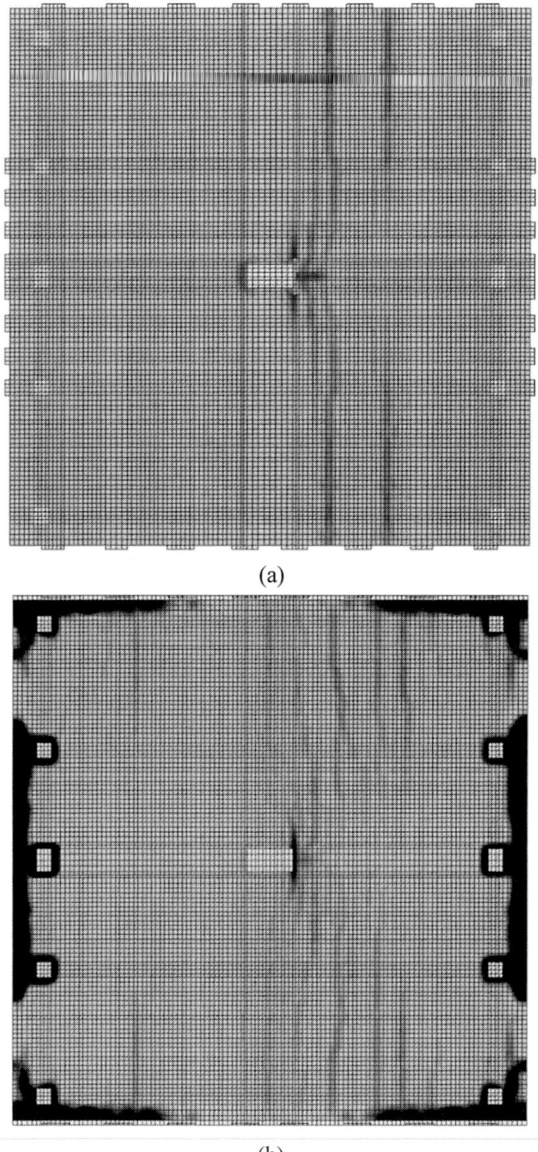

(a)

(b)

Figure 5.48 Crack pattern from bottom view; (a) bonded specimen and (b) unbonded specimen

Since the experimental test was under cyclic loading, the damage pattern could be different from a simulation under monotonic loading. The damage patterns of two simulations are compared (Figures 5.46, 5.47 and 5.48). It is observed that even the global responses from two simulations have negligible differences; the local damage patterns are not the same (flexural cracks). The damaged area (black area) at the periphery of the unbonded slab is mainly tension splitting crack due to lack of anchorage reinforcement.

Figures 5.49 and 5.50 present the strain change distributions of a top mild steel bar and a post-tensioning tendon placed near the column center at several locations and drift levels, both from the experimental data and numerical simulations. The comparisons between the experimental and numerical results indicate a reasonable agreement. The trends in strain change distributions along the length (in the longitudinal direction) are quite well reproduced. The discrepancies might be caused by the different loading conditions during testing (cyclic loading). Also, the higher simulated strain at the drift ratio of 2% compared with the measured value could be due to a perfect bonding assumption, which might not be the case at the ultimate stage. Overall, based on the global and local data comparisons, it is demonstrated that the contact formulation along with other proposed modeling schemes are capable of modeling grout-bonded PT two-way slabs and slab-column connections reasonably well. The experimental and numerical results can be used for better understanding the difference in behavior between the two systems and development of related provisions of the code as described in the next section.

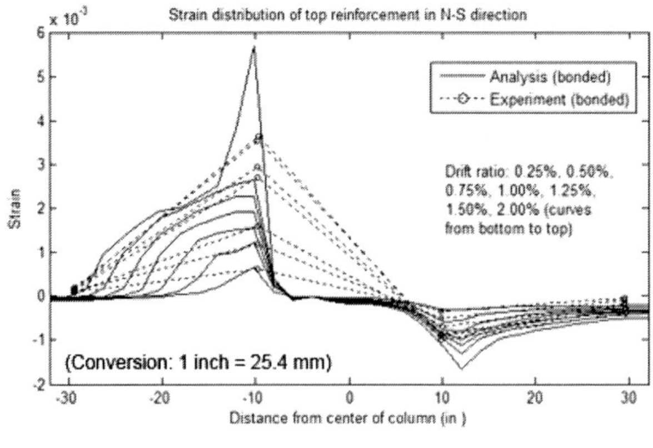

Figure 5.49 Simulated and measured strains in top bonded mild steel (Prawatwong et al., 2007)

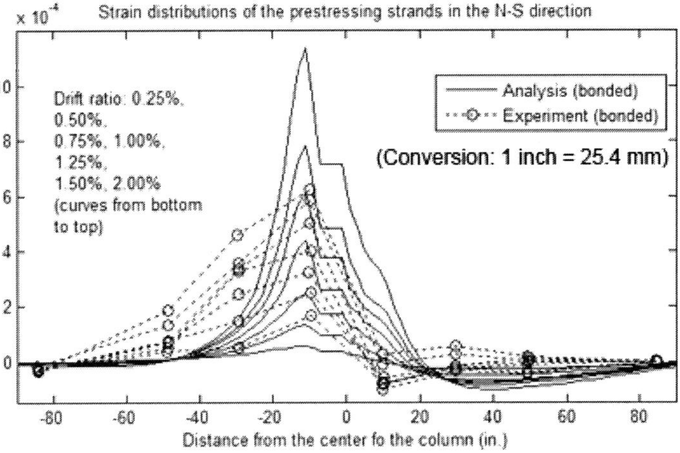

Figure 5.50 Simulated and measured strains in grout-bonded PT tendon (Prawatwong et al., 2007)

5.4.4 Assessments of ACI 318-19 punching shear provisions and comparative study of bonding influence

ACI 318-19 (2019) code provides an empiric model (the so-called eccentric shear stress model) dealing with moment shear interactions at the slab-column connections. This model was described in the previous numerical studies. The shear stresses on the critical section for an interior slab-column connection are assumed as a summation of direct shear and eccentric shear due to unbalanced moment transfer. Figure 5.51 shows the assumed shear stresses on the critical section of interior slab-column connections, as follows

$$V_u = \frac{V_g}{A_c} + \frac{\gamma_v M_u C}{J_c} \tag{5.15}$$

where V_g is the column reaction force; A_c is the area of the shear critical section specified in ACI 318-19; C is the distance from the central line of the critical section (parallel to the axis of unbalanced moment) to the location where the shear stress is calculated; and J_c is the polar moment of inertia of the critical section.

In order to assess the unbalanced moment transfer factor γ_v according to Equation 5.15, the shear stresses, V_u, are extracted from four corner points of the critical section. The shear stresses of two points located on the same side (either the north or south side) of the critical section are identical to each other according to the eccentric shear stress model. Therefore, shear stresses from two points are averaged to reflect the design shear stress on each side (both sides parallel to the axis of unbalanced moment). Furthermore, the shear stresses are assumed to be the same along the slab thickness direction in accordance with the eccentric shear stress model. The shear stresses extracted from five integration points of the simulations are further averaged. The shear stress extraction scheme is shown in Figure 5.51(a).

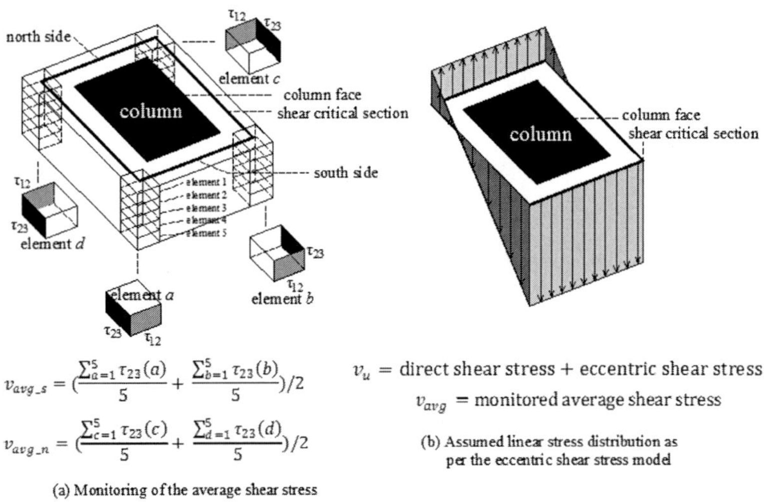

Figure 5.51 The eccentric shear stress model and corresponding numerical shear stress extraction scheme for an interior slab-column connection

Figure 5.52 shows the processed shear stress extractions from both bonded and unbonded simulations. It should be noticed that the shear stresses on the south side have negative values (downward shear stress is assumed to be positive); absolute values are presented in Figure 5.52 for the purpose of comparison. It is apparent that negligible difference exists between the bonded and unbonded condition in terms of shear redistribution. Both cases actually present similar shear stress conditions along the critical section. Following the aforementioned extraction scheme (averaging stresses from five layers), the shear stresses around the critical section are plotted at different drift ratios for both cases as shown in

Figure 5.53. The initial shear stresses at a drift ratio of 0% are identical due to the fact that both bonded and unbonded slab were unbonded during prestressing. Along with a drift ratio up to 2%, slight differences are noticed which again indicates that the shear stresses along the critical section are hardly influenced by the bonding condition of prestressing tendons under similar two-way PT slab designs.

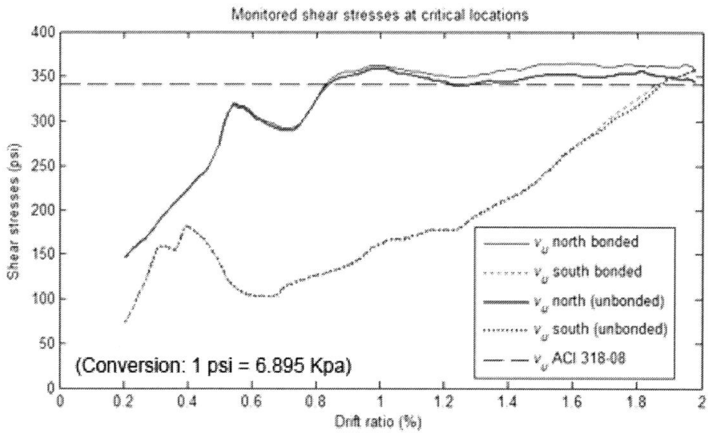

Figure 5.52 Numerical shear redistribution at north side of the critical section from both simulations (bonded vs. unbonded)

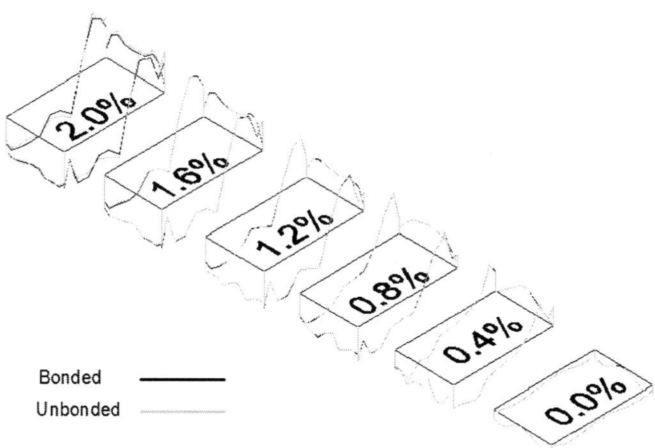

Figure 5.53 3-D plot of numerical shear stress along critical section from both simulations (bonded vs. unbonded)

Following the previous procedures, the unbalanced moment transfer factor, γ_v was obtained via Equation 5.15. Figure 5.54 shows γ_v derived at the north and south sides separately from both bonded and unbonded simulations where negligible differences are observed. Very similar moment shear interactions at slab-column interior connections are suggested for both bonded and unbonded conditions. It is noticed that only one γ_v value is adopted in the design shear stress calculation as per the eccentric shear stress model, while γ_v derived at the south side was slightly larger than that at the north side from the simulation. It could be explained that the north side suffered the most damages when the lateral load was applied which led to a smaller moment shear interaction on this side. The less damaged south side, therefore, shows a higher unbalanced moment transfer ratio. In addition, both the shear capacity and unbalanced moment transfer ratio computed from ACI 318-19 show a reasonable agreement with the numerical results which are not the case in exterior and corner slab-column connections studied previously. Therefore, the suggestion of relief of unbalanced moment transfer ratio might not need to be applied to interior slab-column connections with a similar design.

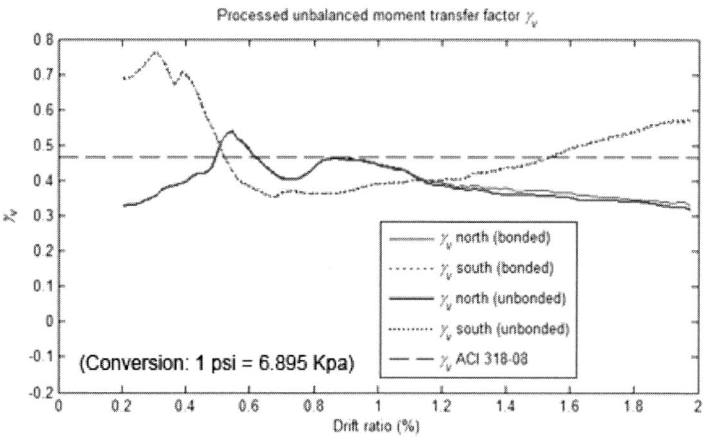

Figure 5.54 Numerical derived unbalanced moment transfer ratio from both simulations (bonded vs. unbonded)

5.4.5 Conclusion

An investigation of the influence of PT bonding conditions on the moment transfer mechanism at a PT interior slab-column connection was attempted using finite element modeling techniques. The direct contact formulation was employed to simulate the interaction between grout-bonded or unbonded tendons and sheathings. The modeling procedures were validated through numerical simulations against the well-documented experimental data. Very good agreement between the simulation and experiment was shown. The following conclusions were drawn based on this study:

(1) The flexural strength and behavior of the two-way bonded PT slab with an interior slab-column connection appeared to be marginally influenced by the bonding condition of PT tendons.

(2) The effect of the PT bonding condition on the moment transfer mechanism of a PT two-way slab-interior column connection also was apparently negligible prior to punching shear failures.

(3) The unbalanced moment transfer ratio and shear capacity obtained from the numerical simulations comply well with ACI 318-19 provisions.

(4) Further studies of the interior slab-column connection are needed to derive a general recommendation.

5.5 Numerical Simulations of Six One-Way PT Slabs and Three PT Beams

5.5.1 Description of specimens

PT One-way Beams with Different PT Bonding Conditions

Mattock et al. (1971) conducted an experimental study of seven simply supported and three continuous beams. The beams are categorized into four groups: T-beams (CB1, CU1, and CU2), continuous over two spans of 28 ft each; simply supported T-beams (TB1, TU1, and TU2) with 28 ft spans; simply supported rectangular-section beams (RB1, RU1, and RU2) with 28 ft spans; and a simply supported T-beam (TU3) identical to TU1 or TU2, except for the amount of bottom mild steel and the presence of a single 3/8 in. diameter non-prestressed seven-wire strand provided as additional bonded bottom reinforcement. Specimens CB1, TB1, and RB1 contained grout-bonded PT tendons inside the corrugated ducts, and they are counterparts of CU1, TU1, and RU1 with unbonded PT tendons, respectively.

Each of them was prestressed by two 1/2 in. Grade 270 seven-wire strands. The tendons were draped parabolically in all simply supported beams with an effective depth of 10 in. at mid-span and zero eccentricity at the ends of the beam, except for RB1, RU1, and RU2 with 2 in. eccentricity at the end. In all continuous beams with two spans, the tendons were draped parabolically with an effective depth of 10 in. at both mid-span locations and zero eccentricity at the ends of the beams.

In all beams, two #2 bars were placed at the top and bottom of the beam throughout the beam length, respectively. For CU1 and TU1, two additional #4 bars and one #2 bar were placed in the same layer of bottom bonded reinforcement with a length of 20 ft, centered on the mid-span, while for CU2/TU2 and RU1/RU2, two #2 bars and a combination of two #3 and one #4 bars were additionally provided, respectively. Also, two #3 and one #4 bars each 20 ft long were provided at the top of the positive moment regions of continuous beams (CU1 and CU2), respectively.

The ultimate strength of the strands was 280 ksi, and the stress at one percent elongation was 255 ksi. The yield strengths of #2, #3 and #4 bars were 54.7, 50 and 46.7 ksi, respectively. The concrete compressive strength was about 4 ksi at testing. As all specimens were half-scaled, concrete blocks weighing up to 100% of the self-weight of the scaled beam were added. Each span was then subjected to four equal point external loads, which were applied at 1.5 and 5.5 ft. away from the mid-span on either side, monotonically, until the failure of the test beams. For the four point loading setup, a spreader beam was used. Details of reinforcement for Specimens RB1, RU1, and RU2 are shown in Figure 5.55. The three specimens were modeled in finite element packages for further comparative research of bonding influence.

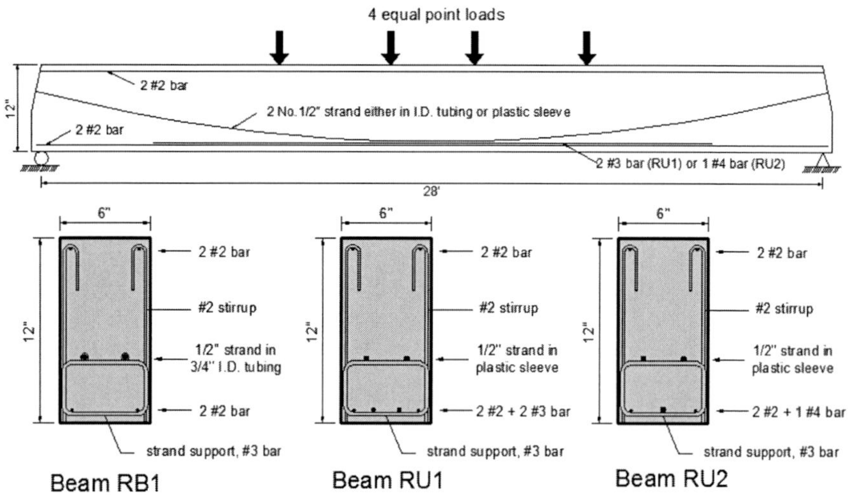

Figure 5.55 Reinforcing details of grout-bonded and unbonded PT beams; experiments conducted by Mattock et al. (1971)

PT One-way Slabs with Different PT Bonding Conditions

Cooke et al. (1981) tested twelve simply supported one-way PT slabs under two-point external loads. Nine of them were prestressed with unbonded tendons and the rest were prestressed with grout-bonded tendons. Of the nine unbonded slabs, three (Slab1, Slab2, and Slab3) had a length of 15 ft 9 in. with a span length of 15 ft 1-1/8 in.; three (Slab4, Slab5, and Slab6) had a length of 11 ft 9-3/4 in. with a span length of 11 ft 1-3/4 in.; and three (Slab7, Slab8, and Slab9) had a length of 7 ft 10-1/2 in. with a span length of 7 ft 2-5/8 in. The three bonded slabs (SlabB4, SlabB5, and SlabB6) were identical to Slab4, Slab5, and Slab6, respectively, except the tendons were bonded to the concrete. The slabs in each unbonded group had transverse widths of 1 ft 1-7/8 in., 2 ft 3-3/4 in. and 3 ft 10-1/2 in., respectively. All slabs were 7-1/16 in. thick. The first two slabs in each group were prestressed with three straight tendons of 1/2 in. diameter, whereas the last slab in each group had three straight tendons of 5/8 in. diameter. All tendons were placed at an eccentricity of 4-3/4 in. No bonded reinforcement was provided to any of the slabs.

The target concrete compressive strength was 5 ksi for all slabs. The ultimate tensile strength for 1/2 in. and 5/8 in. tendons were 267 ksi and 256 ksi, respectively. All slabs were subjected to line loads applied at two points located at 4 ft 11-1/16 in. (Slab1, Slab2, and Slab3), 3 ft 7-5/16 in. (Slab4, Slab5, Slab6, SlabB4, SlabB5, and SlabB6) or 2 ft 3-9/16 in. (Slab7, Slab8, and Slab9) from each support center. The specimen was statically loaded in increments until failures occurred. Slabs4, Slab5, Slab6, SlabB4, SlabB5, and SlabB6 shown in Figure 5.56 were modeled and simulated for further studies.

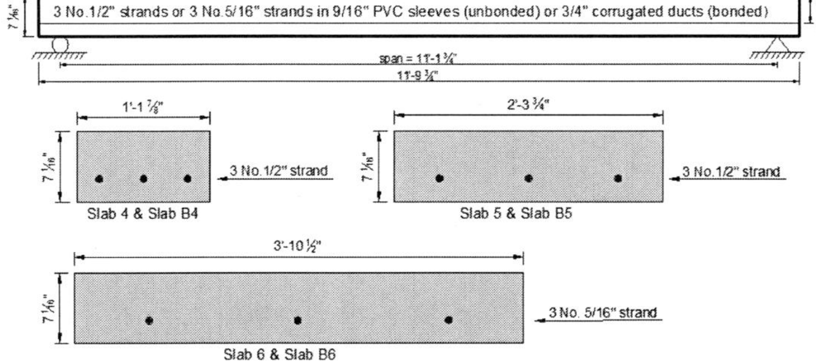

Figure 5.56 Reinforcing details of grout-bonded and unbonded PT one-way slabs; experiments conducted by Cooke et al. (1981)

5.5.2 Numerical models

Due to the presence of grout-bonded specimens, the PT system either with unbonded or bonded conditions was modeled by the contact formulation as employed in the previous study. In addition, two unbonded PT beams of RU1 and RU2 and two unbonded PT one-way slabs of Slab4 and Slab5 were also modeled by the spring system method. Other modeling schemes were followed the same as previous numerical simulations. A typical finite element mesh for the one-way slabs and beams is shown in Figure 5.57.

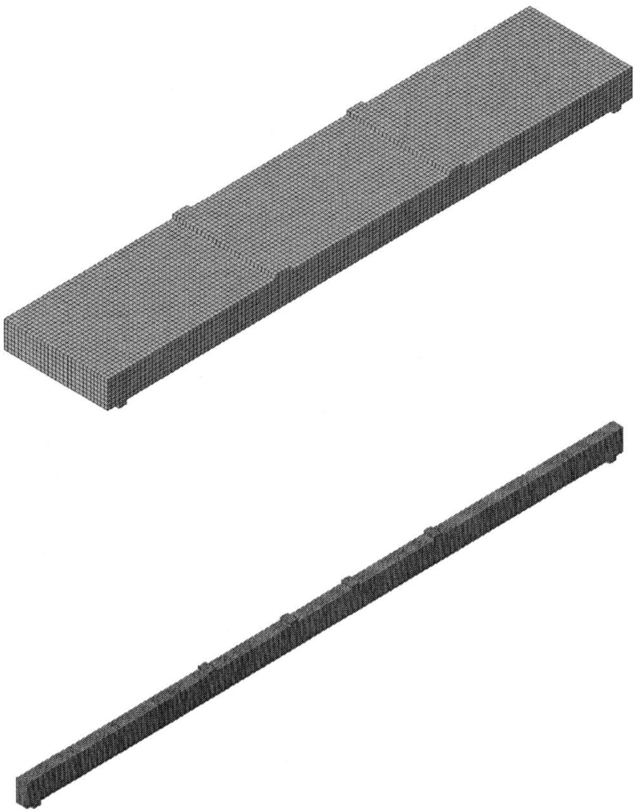

Figure 5.57 Finite element mesh used for one-way PT slabs and simply supported beams

5.5.3 Numerical results and validations

Comparisons of applied load versus mid-span deflection were made between numerical simulations and experiments. The results are shown in Figure 5.58. All simulations have a good agreement with experiments. Specimens RU1 and RU2 were modeled by both contact formulation and spring method, and their corresponding numerical simulations showed that both approaches are valid. Figure 5.58 showing negligible difference between the two approaches and experiments, validates that both approaches are appropriate for simulating unbonded PT tendons in one-way PT members. Damage patterns are also compared with available experimental data (Figures 5.59, 5.60, and 5.61). Note that only simulations with the contact formulation are shown and compared. The numerical damage patterns show a good agreement with experiments. Particularly, the simulation with a bonded condition tends to develop cracks more uniformly along the beam than unbonded specimens which corresponds with experimental observations in practice.

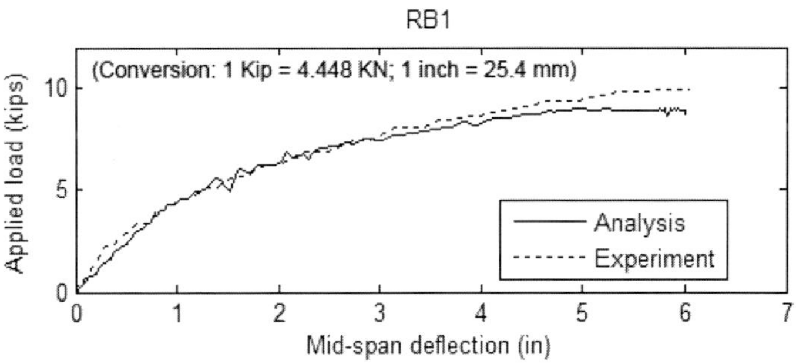

(a)

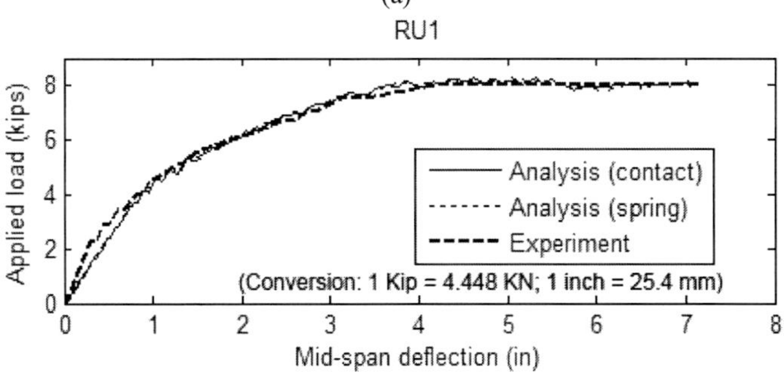

(b)

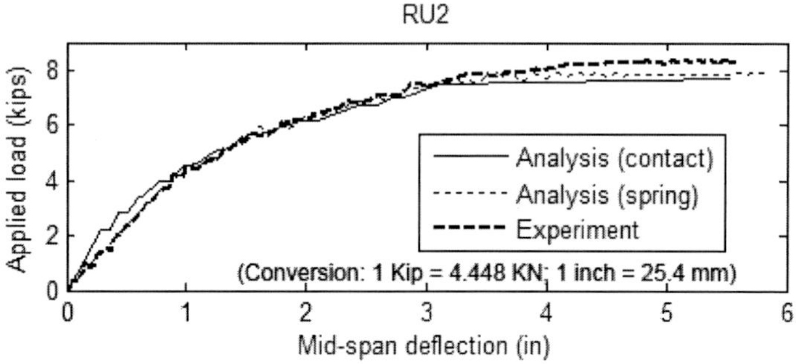

(c)

Figure 5.58 Comparisons of global responses between numerical simulations and experiments for unbonded PT beams; experiments conducted by Mattock et al. (1971)

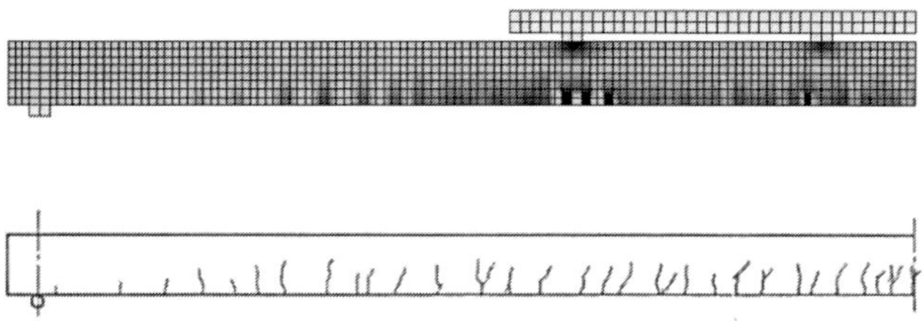

Figure 5.59 Damage patterns of RB1; simulation with contact formulation (top) and experiment (bottom)

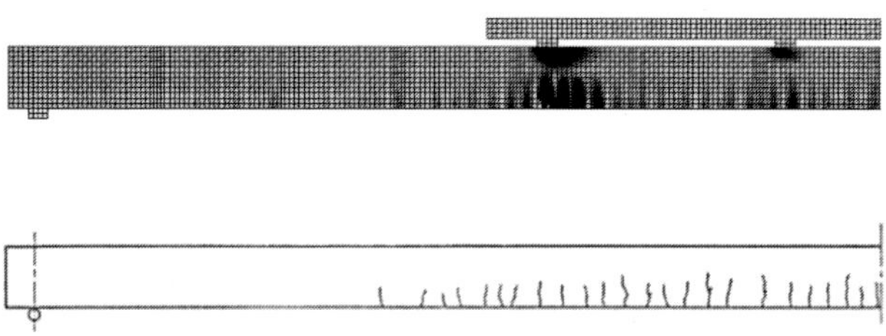

Figure 5.60 Damage patterns of RU1; simulation with contact formulation (top) and experiment (bottom)

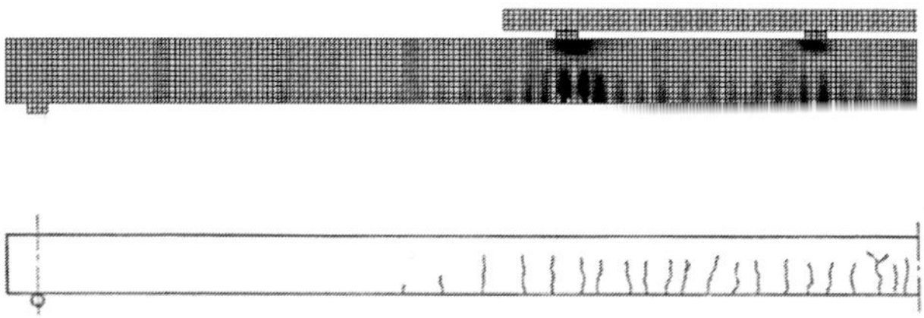

Figure 5.61 Damage patterns of RU2; simulation with contact formulation (top) and experiment (bottom)

The global responses of all unbonded slab and bonded slab simulations are shown in Figures 5.62 and 5.63, respectively. All simulations have good agreements with experiments including the damage patterns at ultimate stage (Slab6 and SlabB6 are extremely low reinforced, and the significant fluctuations of applied loads after brittle cracking is a nature of explicit dynamic analysis procedure). Specimens Slab4 and Slab5 were also modeled by the spring method. The global responses exhibit very similar to the modeling results based on the contact formulation. The damage pattern at the constant moment region from both simulations (contact formulation) and experiments are shown in Figures 5.64 to 5.69. Generally, the patterns shown by numerical simulations are in good agreement with the corresponding experiments. All flexural cracks were developed at the constant moment regions, corresponding with experiments very well.

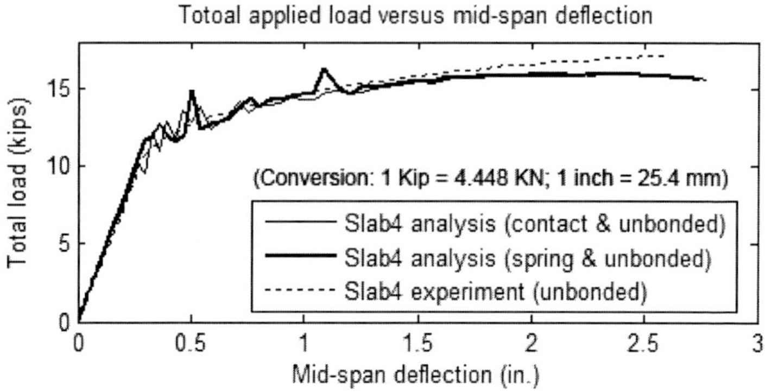

(a)

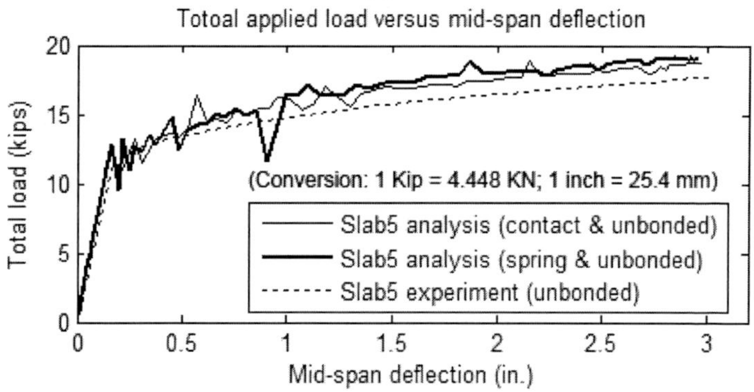

(b)

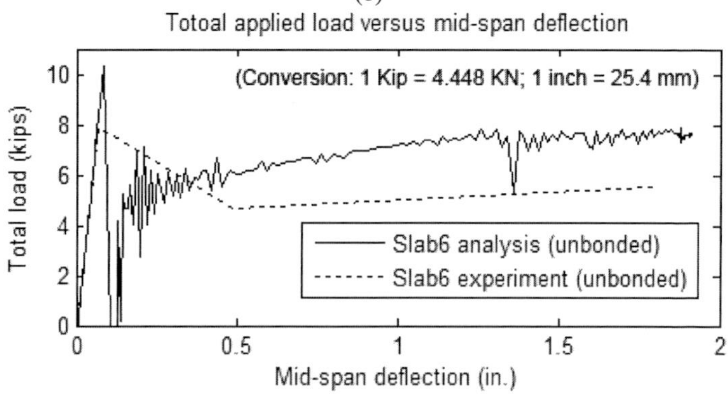

(c)

Figure 5.62 Comparisons of global responses between numerical simulations and experiments for unbonded PT one-way slabs; experiments conducted by Cooke et al. (1981)

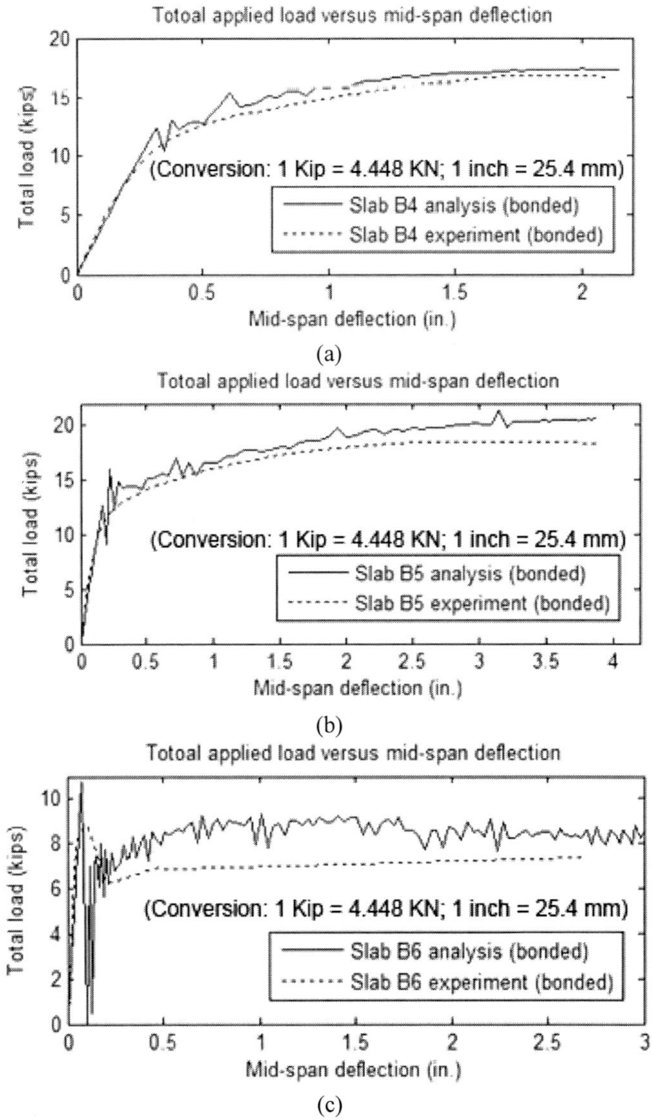

Figure 5.63 Comparisons of global responses between numerical simulations and experiments for grout-bonded PT one-way slabs; experiments conducted by Cooke et al. (1981)

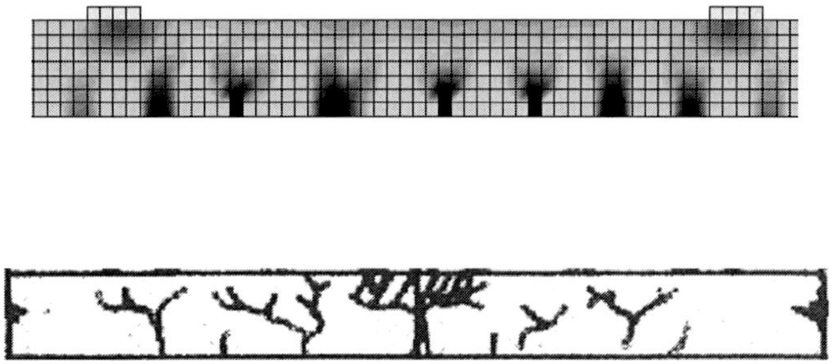

Figure 5.64 Damage patterns at constant moment region of Slab4; simulation with contact formulation (top) and experiment (bottom)

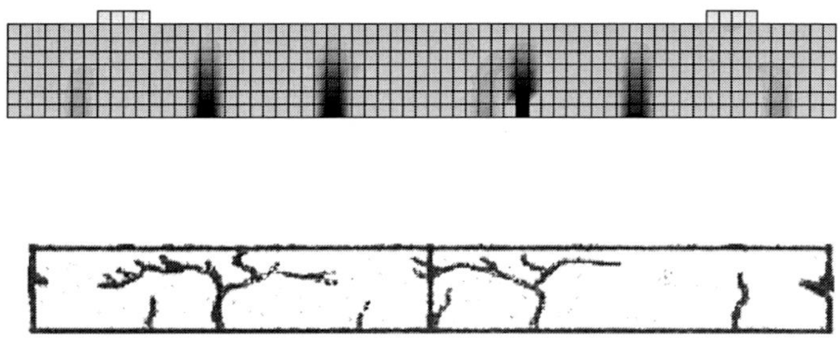

Figure 5.65 Damage patterns at constant moment region of Slab5; simulation with contact formulation (top) and experiment (bottom)

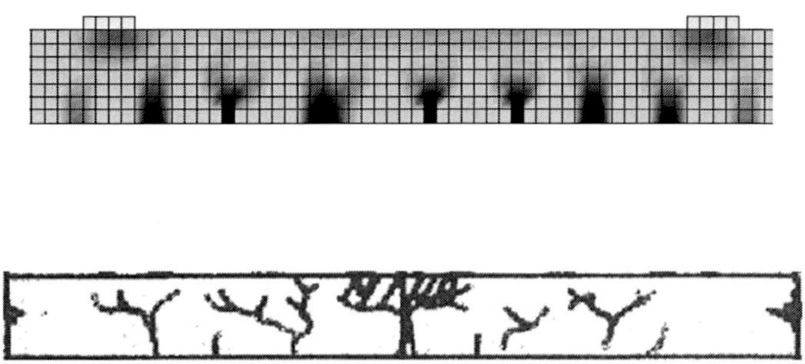

Figure 5.66 Damage patterns at constant moment region of Slab6; simulation with contact formulation (top) and experiment (bottom)

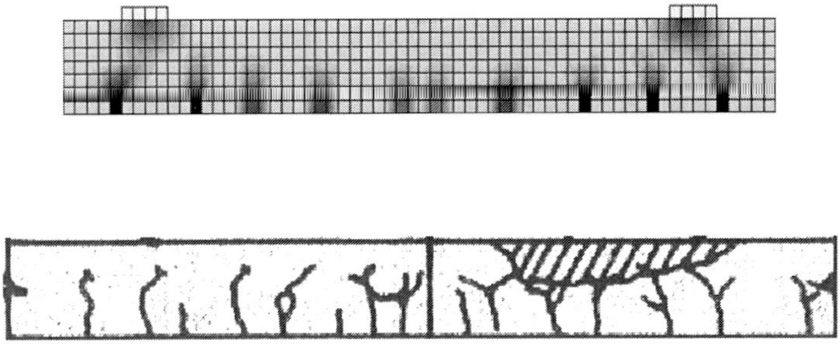

Figure 5.67 Damage patterns at constant moment region of SlabB4; simulation with contact formulation (top) and experiment (bottom)

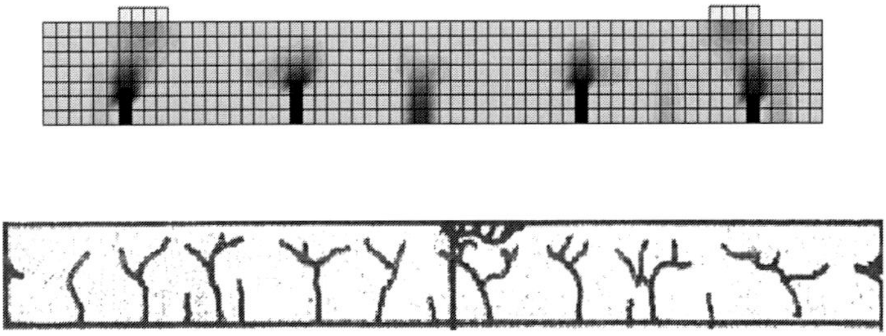

Figure 5.68 Damage patterns at constant moment region of SlabB5; simulation with contact formulation (top) and experiment (bottom)

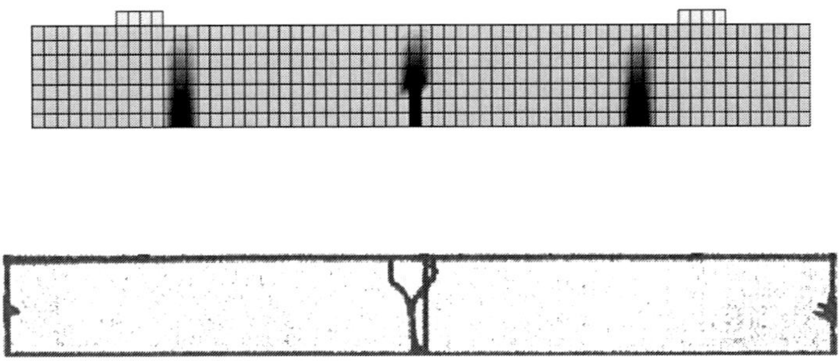

Figure 5.69 Damage patterns at constant moment region of SlabB6; simulation with contact formulation (top) and experiment (bottom)

Table 5.5 summarizes the applied load of each specimen at experimental termination point of mid-span deflection.

Table 5.5 Summary of applied load (kips) at experimental termination point of mid-span deflection

	FEA	Experiment	FEA/Expt
RB1	8.93	9.9	0.9
RU1†	8.06	8.02	1.005
RU2†	7.83	7.7	1.02
Slab4†	15.85	17.07	0.93
Slab5†	18.7	17.8	1.05
Slab6‡	10.32	7.87	1.31
SlabB4	17.34	16.64	1.04
SlabB5	20.58	18.31	1.12
SlabB6‡	10.77	8.85	1.21

† Data obtained from contact formulation
‡ Applied load obtained at cracking point

5.5.4 Comparative study of tendon stress and moment capacity at ultimate

The ultimate stress in tendons was monitored as well by comparison with experimental and theoretical calculations. For unbonded specimens, the stress is nearly uniformly distributed along the tendon due to a frictionless assumption. The ultimate tendon stress was read at the integration point where the largest stress appears from both bonded and unbonded simulations. The ultimate tendon stress f_{ps} obtained by ACI 318, Yang and Kang's method (Yang and Kang, 2011), numerical readings as experimental data are presented in Table 5.6. ACI 318-19 permits the ultimate tendon stress in an unbonded PT member with span-to-depth ratio of 35 or less to be calculated as:

$$f_{ps} = f_{se} + 10000 + \frac{f_c'}{100\rho_p} \tag{5.16}$$

where f_{ps} should not be taken greater than the less of f_y and $f_{se} + 60000$

The Yang and Kang's method predicts f_{ps} as:

$$f_{ps} = \frac{-B_1 + \sqrt{B_1{}^2 - 4A_1C_1}}{2A_1} \leq f_{py} \tag{5.17}$$

where A_1 , B_1 and A_1 are coefficients determined by reinforcement configurations, concrete and steel material properties as well as loading types. Details can be referred to Yang and Kang (2011).

ACI provides an equation to calculate ultimate tendon stress for members with bonded tendons as:

$$f_{ps} = f_{pu}\left\{1 - \frac{\gamma_p}{\beta_1}\left[\rho_p \frac{f_{pu}}{f_c'} + \frac{d}{d_p}(\omega - \omega')\right]\right\} \tag{5.18}$$

where A_1 is $\rho\, f_y/f'_c$, ω' is $\rho'\, f_y/f'_c$, and γ_p is 0.55 for f_{py}/f_{pu} not less than 0.80; 0.40 for f_{py}/f_{pu} not less than 0.85; 0.28 for f_{py}/f_{pu} not less than 0.90. If any compression reinforcement is taken into account when calculating f_{ps} by Equation 5.18, the term

$$\left[\rho_p \frac{f_{pu}}{f_c'} + \frac{d}{d_p}(\omega - \omega')\right]$$

shall be taken not less than 0.17 and d' shall be no greater than $0.15d_p$.

According to Equations 5.16 and 5.17, ACI 318-19 predicts the closest f_{ps} to experimental data comparing with numerical simulations and Yang and Kang's method for unbonded PT beams of RU1 and RU2. The similar moment capacities of RU1 and RU2 are due to the similar configurations of reinforcement. It is noted that the corresponding moment capacity M_n of RB1 obtained from both ACI 318 and numerical simulations are smaller than the test. It was explained by Mattock et al. (1971) that the steel tube used for bonded prestressing tendon acted as additional bonded reinforcement. Given the fact that there was more bonded reinforcement in RB1 than RU1 and RU2 in the test, similar moment strengths could be expected for a beam prestressed with bonded and unbonded tendons under the same amount of bonded reinforcement.

As shown in Table 5.6, ACI 318-19 predictions of f_{ps} for unbonded Specimens Slab4 and Slab5 are the lowest compared with data obtained from experimental and other methods. Because the extremely low reinforcement configuration of Slab6 is an unlikely practical design, ACI

318-19 tends to overestimate the ultimate stress while the finite element analysis underestimates it. On the other hand, Yang and Kang's method cannot be applied to predict ultimate stress for Slab6 due to the extremely low reinforcement ratio. The ultimate tendon stresses in bonded Specimens SlabB4, SlabB5 and SlabB6 predicted by ACI 318-19 and finite element simulations are similar. The slightly higher stresses from finite element simulations could result from the assumption of perfect bonding in the modeling. The moment strengths calculated according to corresponding f_{ps} are close to experiments for bonded specimens. It is obvious that the experimental moment capacities of Slab4 and SlabB4, Slab5 and SlabB5 are similar. Even though the moment capacities calculated according to f_{ps} of finite element simulations are slightly larger than their counterparts in bonded cases, the similar applied loads of specimens in pairs (Figures 5.62 and 5.63) reveal that bonding of prestressing tendons is not a significant factor influencing moment capacity. The pair of Slab6 and SlabB6 is the only exception which has a relatively large difference in moment capacity. The reason is simply that the moment capacities of Slab6 and SlabB6 were controlled by the cracking moment due to extremely low reinforcement. For bonded Specimen SlabB6, the bonding between prestressing tendons and concrete resulted in a higher tendon stress than an unbonded one when dominant cracking occurred. Consequently, the cracking moment of SlabB6 was larger than Slab6. Similar moment strength could be expected if additional bonded reinforcement is added to both slabs.

Table 5.6 Summary of tendon stress and moment capacity at ultimate stage for each specimen

Specimens			RB1	RU1	RU2
fpe (Expt)	ksi		188.2	183.1	186.6
fps (Expt)	ksi		NA	208	205
fps (ACI 318)	ksi		250.7	202.5	206
fps (Yang & Kang)	ksi		NA	212.3	216.3
fps (Analysis)	ksi		233.1	214 *	212 *
Mn ‡	(Expt)	ksi	827	707	689
Mn †	(ACI 318)	in.-kips	708.5	655	650
Mn †	(Yang & Kang)	in.-kips	NA	674.9	669.3
Mn †	(Analysis)	in.-kips	672.1	677.8	661.8

Specimens			Slab4	Slab5	Slab6
fpe (Expt)	ksi		169	167	177
fps (Expt)	ksi		200	209	228
fps (ACI 318)	ksi		190.3	199.6	237
fps (Yang & Kang)	ksi		215.5	229.4	NA
fps (Analysis)	ksi		210.5 *	223 *	220.8 *
Mn ‡	(Expt)	ksi	390	422.4	184.8
Mn †	(ACI 318)	in.-kips	333.1	378	198
Mn †	(Yang & Kang)	in.-kips	368.6	429	NA
Mn †	(Analysis)	in.-kips	361.7	417	184.8

Specimens			SlabB 4	SlabB 5	SlabB 6
fpe (Expt)	ksi		174	169	180
fps (Expt)	ksi		NA	NA	NA
fps (ACI 318)	ksi		226.9	241.5	263.2
fps (Yang & Kang)	ksi		NA	NA	NA
fps (Analysis)	ksi		227.8	248.1	273
Mn ‡	(Expt)	ksi	382.8	441.6	224.4
Mn †	(ACI 318)	in.-kips	382.5	448.6	219.9
Mn †	(Yang & Kang)	in.-kips	NA	NA	NA
Mn †	(Analysis)	in.-kips	383.8	459.5	227.9

† M_n is calculated based on corresponding f_{ps}

‡ M_n is obtained from experimental data

* Obtained from analyses with contact formulation

5.5.5 Conclusion

The influence of bonding conditions of prestressing tendons on flexural strength of beams and one-way slabs was studied through finite element modeling. Direct contact formulation was employed in simulating the interaction between either bonded or unbonded tendons and sheathings. In addition, unbonded PT systems were also attempted to be modeled via the spring system method. Both modeling approaches were validated by several numerical simulations against documented experiments including members prestressed with either bonded or unbonded tendons. Good agreement between all simulations and experiments was observed. The primary variable investigated in this study is the bonding of prestressing tendons which can become either bonded or unbonded after prestressing. Particularly, the flexural strength of beams and one-way slabs with bonded or unbonded tendons was compared through numerical studies of documented tests. Furthermore, the tendon stresses at ultimate computed from various methods were summarized and compared to experimental data. Based on the numerical simulations, the following conclusions can be drawn:

(1) The flexural strengths of the beams and one-way slabs are hardly influenced by the bonding condition of prestressing tendons.

(2) The current finite element modeling schemes yield a slightly overestimated tendon stress at ultimate. This might result from the neglecting prestress loss and the assumption of uniformly distributed initial strain field of tendon due to prestress.

(3) Future projects are desired towards improving the modeling schemes of PT tendon systems which are able to incorporate the post-tensioning procedures in practice.

5.6 Summary

This chapter presented the validation of the proposed numerical modeling schemes and the application to various engineering problems in practice such as the moment shear interaction of slab-column connections in two-way PT slabs, the PT structural performance under different PT tendon bonding conditions and the tendon stress at ultimate state. Overall, the proposed modeling schemes performed very well compared to documented experimental data and served as a powerful tool to assist solving complicated practical problems. Particularly, the moment shear interaction was investigated through four exterior connections, two corner connections and one interior connection. The punching shear provisions were heavily assessed and evaluated in terms of ACI 318-19 through numerous numerical simulations along with the experimental data. The conclusion drawn from the series of studies of ACI 318-19 is conservative of punching shear design with respect to exterior and corner connections in two-way PT slabs. A relief of the unbalanced moment transfer ratio is suggested and several modifications of the eccentric shear stress model are recommended according to the analyses. The comparative studies of PT tendon bonding influence on structural performance led to the conclusion that the effect is negligible in several types of structures (e.g., two-way PT slab, one-way PT slab and PT beam). The numerical simulations clearly indicated that both punching shear and flexural behaviors were barely affected by bonding conditions in these structural members. Even though the current modeling schemes based on general purpose finite element packages reproduced structural behavior very well, several limitations were noticed in terms of flexible modeling and reliable results. For instance, the current modeling scheme tends to overestimate stresses in PT tendons at ultimate state. The overestimation might result from an oversimplification of prestressing procedures where prestress loss is totally neglected. This becomes the motivation or developing more flexible and specialized finite element formulation towards simulating PT systems in the next chapter. Chapter 6 presents a robust and general but very flexible finite element frame work for modeling PT frame systems in bonded, partially bonded and unbonded conditions.

CHAPTER SIX

A NONLINEAR FINITE ELEMENT FORMULATION FOR PT STRUCTURES

6.1 Introduction

Chapter 4 described the modeling schemes applied to PT prestressed members via general purpose finite element packages. Based on the proposed methodology, a series of numerical studies were carried out and validated the modeling against experiments in Chapter 5. The performance and robustness of the proposed modeling schemes were justified. However, slightly inaccurate predictions of the local behavior of prestressing tendons were observed in the numerical studies and discussed in the last part of Chapter 5. Again, the simplification of PT prestressing procedure in the simulation might be the reason behind the discrepancy. The previously proposed modeling scheme assumes an uniformly distributed strain field of PT tendons which is caused by prestressing. Additionally, the short-term prestressing loss due to the effect of curved profile, tendon wobble and anchorage wedge seating loss are not addressed. To investigate local behaviors of prestressing tendons, such characteristics should be accounted for in numerical simulations. Therefore, a more flexible and specialized finite element formulation is proposed in this chapter.

The general frame work is shown in Figure 6.1. The proposed finite element formulation accounts for material nonlinearity (concrete crushing and cracking) and steel yielding (Section 6.5). The boundary nonlinearity introduced from interface contact between tendons and corresponding sheathings is formulated through a modified node-to-segment (NTS) contact elements (Section 6.2). The contact model is flexible enough to simulate perfectly unbonded, partially unbonded and bonded interface via appropriate constitutive relations of contact interfaces. The assembly of contact elements is automatically made through a contact search algorithm. Perfectly bonded mild steel bars are constrained by introducing embedding elements (Section 6.3), and the PT anchorage system is modeled through a special purpose element (Section 6.4). A nonlinear

Euler-Bernoulli beam element is implemented for investigating slender PT beams in this study (Section 6.5). The elements proposed and developed are mostly engineering type which considerably saves computational powers. The Newton-Raphson method with line search technique is employed to solve nonlinear equations numerically (Section 6.6). Although the works presented in this chapter describe only a two-dimensional finite element formulation and its application is limited to PT beams with a large span-to-depth ratio, the general framework of the formulation is clearly stated and can be extended to other applications with trivial effort (e.g., analysis of deep beam can be approached by replacing the Euler-Bernoulli beam with Timoshenko beam element). Figure 6.2 illustrates the modeling of a PT frame member through the proposed formulation.

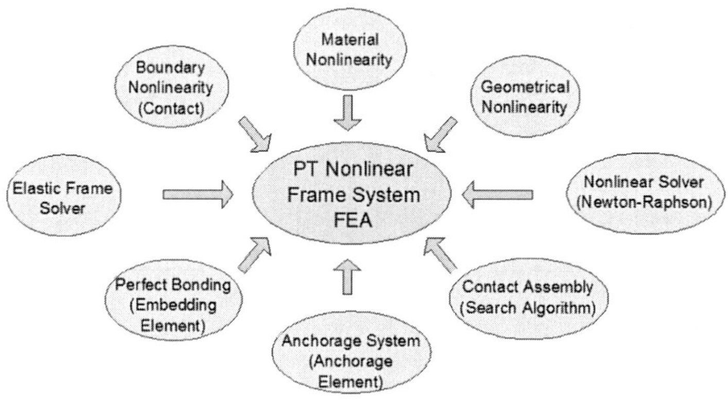

Figure 6.1 General frame work of the proposed finite element formulation

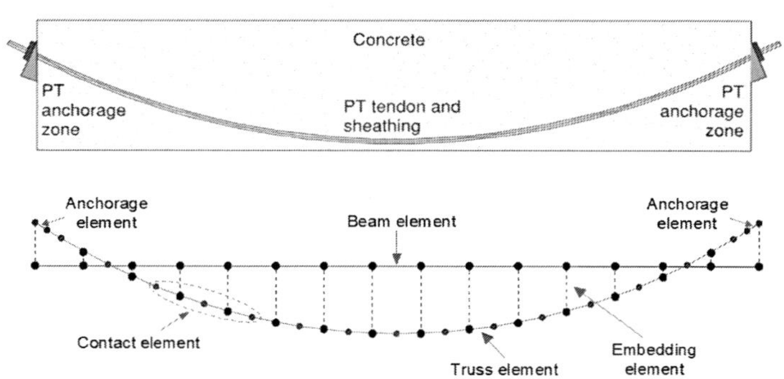

Figure 6.2 Illustration of the finite element discretization scheme

6.2 Contact Element

6.2.1 Node-to-segment discretization

The following NTS based formulation is a modification and extension to structural analysis of the original work done by Wriggers (2008). The slip of tendons can be quite large at prestressing jacking and service stage. Accordingly, a contact discretization scheme which can handle finite sliding in the contact interface is considered. The most simplified discretization scheme to treat large sliding at the contact interface is the so-called node-to-segment (NTS) formulation. One or several slave nodes are in contact with a master surface. Both the prestressing tendon and corresponding sheathing are discretized to two-dimensional linear truss elements in this study. The nodes laying on the tendon elements are slave nodes and are in contact with the segments of sheathing in this case. A typical contact element contains three slave nodes and two master nodes which form the master surface as shown in Figure 6.3 (the node with position vector X_s , X_3 and X_4 represent the middle slave node and adjacent two slave nodes respectively. The straight line between the nodes with position vectors X_1 and X_2 forms the master surface of one sheathing element in this case).

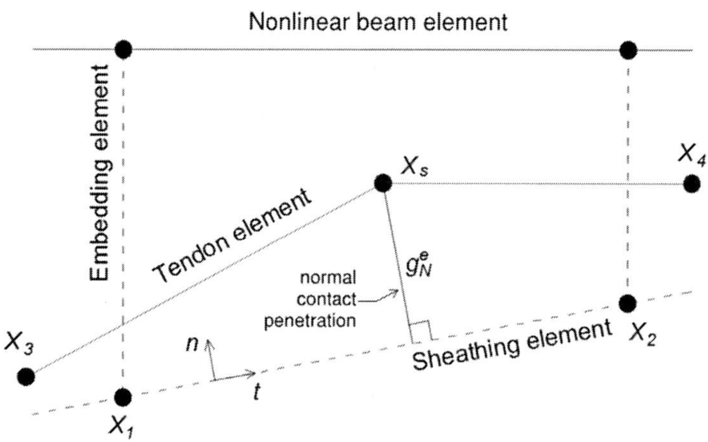

Figure 6.3 Node-to-segment based contact element

The node of prestressing tendon element given by the position vector X_s comes into contact with the master surface formed by sheathing element nodes given by the position vectors X_1 andX_2. The unit normal and unit tangent vectors of the master surface are denoted as n and t, respectively. The deviation distance between the slave node and master surface is defined by a scalar denoted asg_{Ns}, and the finite slip is defined by a scalar denoted asg_{Ts}. With the above definition, the unit tangent vector on the master surface is computed as:

$$t = \frac{(X_2 - X_1)}{l} \tag{6.1}$$

where l is the master segment length defined as:

$$l = \|X_2 - X_1\| \tag{6.2}$$

The unit normal vector is computed as:

$$n = e_3 \times t \tag{6.3}$$

where e_3 is the unit vector in the out of plane direction.

Since linear truss elements are used to formulate tendon sheathing, any point on the master surface is known by a linear interpolation of X_1 and $X_2$with local surface coordinate ξ as:

$$X(\xi) = X_1(1 - \xi) + X_2 \; 0 \leq \xi \leq 1 \tag{6.4}$$

The projection of a slave node onto the master surface is denoted by the local coordinate $\bar{\xi}$ as:

$$\bar{\xi} = \frac{(X_s - X_1) \cdot t}{l} \tag{6.5}$$

The deviation distance g_{Ns} between the slave node and master surface is computed as:

$$g_{Ns} = \left[X_s - \left(1 - \bar{\xi}\right)X_1 - \bar{\xi}X_2\right] \cdot n \tag{6.6}$$

where s is the contact element label.

The first variation of the distance function g_{Ns} is derived as:

$$\delta g_{Ns} = \left[\eta_s - \left(1 - \bar{\xi}\right)\eta_1 - \bar{\xi}\eta_2\right] \cdot n \tag{6.7}$$

where η_s, η_1 and η_2 are the displacement vectors of slave node and master nodes, respectively.

The slip distance function is derived as:

$$g_{Ts} = \left(\bar{\xi} - \xi_0\right) l \tag{6.8}$$

where ξ_0 denotes the local surface coordinate of the slave node projected on the master surface at the initial configuration.

With first variation,

$$\delta g_{Ts} = l\delta\bar{\xi} + \left(\bar{\xi} - \xi_0\right)\delta\, l \tag{6.9}$$

where

$$\begin{gathered} \delta\bar{\xi} = \frac{[\eta_s - (1-\bar{\xi})\eta_1 - \bar{\xi}\eta_2]\cdot t}{l} + \frac{g_{Ns}[\eta_2 - \eta_1]\cdot n}{l^2} \\ \delta\, l = [\eta_2 - \eta_1]\cdot t \end{gathered} \tag{6.10}$$

Substituting Equation 6.10 into Equation 6.9, the first variation of g_{Ts} with respect to nodal displacement increments is derived as:

$$\begin{aligned} \delta g_{Ts} = [\eta_s - (1-\bar{\xi})\eta_1 - \bar{\xi}\eta_2]\cdot t + \frac{g_{Ns}[\eta_2 - \eta_1]\cdot n}{l} \\ + \frac{g_{Ts}[\eta_2 - \eta_1]\cdot t}{l} \end{aligned} \tag{6.11}$$

Equations 6.7 and 6.11 represent the essential kinematic relations of the contact element described in this study. By the finite element discretization, the virtual work from contact contribution is given as:

$$\delta\Pi = \sum_{s=1}^{n_c} (P_{Ns}\delta g_{Ns} + T_{Ts}\delta g_{Ts}) \tag{6.12}$$

where P_{Ns} and T_{Ts} are the normal and tangential contact forces, respectively, and are determined by the constitutive equation chosen at the contact interface. n_c represents the total number of contact elements. S represents the element label of contact elements.

For this study, a hard contact formulation is introduced in the normal direction while a frictional contact associated with Coulomb's law is employed in the tangential direction. The descriptions of two formulations are presented in the later section. The kinematic constraint appearing in the current finite element formulation is enforced by the penalty method in which very large penalty stiffness ε_N and ε_T are selected to compute the normal and tangential contact forces, respectively. The penalty factors are orders of magnitudes larger than typical stiffness in the beam such that the erroneous penetration and slip distances are miniscule. For an ideal bonded tendon system, stick behavior is observed in the tangential direction. Therefore, the contact forces are determined by the penalty method in both the normal and tangential directions. In the case of partially unbonded and perfectly unbonded systems, sliding occurs at the contact interface of the tendon and corresponding sheathing. The tangential contact force is given by the integration of the friction law as described in Section 6.2.3. In the case of the stick, normal and tangential contact forces are given by:

$$\begin{aligned} P_{Ns} &= \varepsilon_N g_{Ns} a_s \\ T_{Ts} &= \varepsilon_T g_{Ts} a_s \end{aligned} \tag{6.13}$$

where a_s is the area of the contact element which is equal to length of the truss element that discretizes the tendon sheathing.

In the case that multiple slave nodes in contact with the same master segment, the contact area a_s is no longer the whole length of the element. Instead, a_s is computed from the midpoints between the projections $\bar{\xi}$ of the neighboring slave nodes as shown in Figure 6.4.

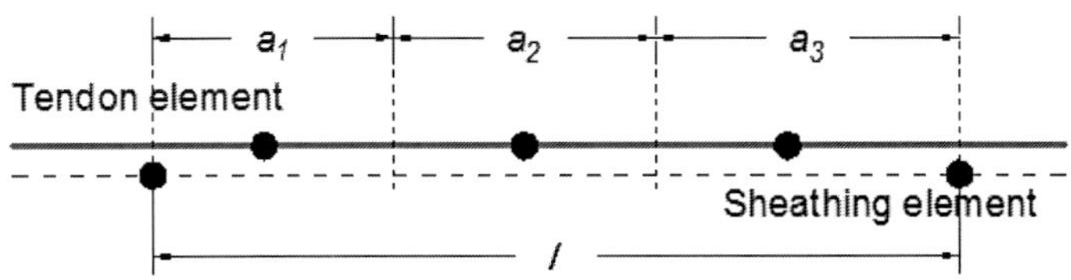

Figure 6.4 Multiple slave nodes on one segment

6.2.2 Contact element residual force

Firstly, the element displacement vector of the contact element is defined as:

$$\bar{\eta} = (\eta_s\ \eta_1\ \eta_2\ \eta_3\ \eta_4)^{\mathrm{T}} \tag{6.14}$$

Even though the tendon sheathing element is discretized into linear truss elements, each node contains the same three degrees-of-freedom as two translational degrees and one rotational degree. This ensures the compatibility of connecting the tendon sheathing to the beam by means of using an embedding element. The embedding element is modeled by the Euler-Bernoulli beam element which shares the nodes of the prestressed beam element and the tendon sheathing element as shown in Figure 6.3. Therefore, the displacement vector contains twelve degrees-of-freedom in which two rotational degrees-of-freedom have nothing to do with the NTS formulation. Accordingly, the N_s, N_{0s}, T_s and T_{0s} matrices are defined as:

$$N_s = \left[n\ 0\ -(1-\bar{\xi})n\ 0\ 0\ -\bar{\xi}n\ 0\ 0\ 0\ 0\ 0\ 0\right]^{\mathrm{T}} \tag{6.15}$$

$$N_{0s} = [0\ 0\ -n\ 0\ 0\ n\ 0\ 0\ 0\ 0\ 0\ 0\ ]^{\mathrm{T}} \tag{6.16}$$

$$T_s = \left[t\ 0\ -(1-\bar{\xi})t\ 0\ 0\ -\bar{\xi}t\ 0\ 0\ 0\ 0\ 0\ 0\right]^{\mathrm{T}} \tag{6.17}$$

$$T_{0s} = [0\ 0\ -t\ 0\ 0\ t\ 0\ 0\ 0\ 0\ 0\ 0\ ]^{\mathrm{T}} \tag{6.18}$$

The first variation of the penetration function is written as:

$$\delta g_{Ns} = \bar{\eta}_s^T N_s \tag{6.19}$$

Using similar notation, the first variation of the tangential sliding function is written as:

$$\delta g_{Ts} = \bar{\eta}_s^T \left(T_s + \frac{g_{Ns}}{l} N_{0s} + \frac{g_{Ts}}{l} T_{0s}\right) = \bar{\eta}_s^T \hat{T}_s \tag{6.20}$$

The discrete weak form of contact contribution in terms of virtual work is described as:

$$\delta g_{Ns} P_{Ns} + \delta g_{Ts} T_{Ts} = \bar{\eta}_s^T \left(N_s P_{Ns} + \hat{T}_s T_{Ts}\right) \tag{6.21}$$

Equation 6.21 yields the contact element residual force vector as:

$$G_s = N_s P_{Ns} + \hat{T}_s T_{Ts} \tag{6.22}$$

However, using Equation 6.22 would lead to inaccurate prestress calculation as illustrated in Figure 6.5. The normal vector of sheathing P_N^e (master segment between X_1 and X_2) and imaginary normal contact force vector $\bar{P}_N^e$ are most likely misaligned due to the discretization scheme. A small correction force term $P^e_{N\_cor}$ to correct the misalignment must be introduced to balance the internal prestress of tendon.

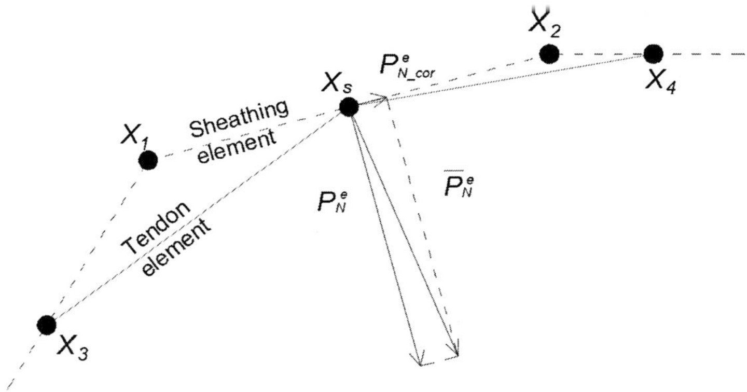

Figure 6.5 Illustration of the tangential correction force to eliminate artificial prestress loss or gain along the tendon path

The correction force term can be computed as:

$$P^e_{N\_cor} = P^e_N |tan(\alpha)| t' \tag{6.23}$$

Where *t'* is the unit direction vector for the correction force; α is the angle between the master segment normal vector and imaginary normal contact force vector derived from following equations:

$$tan(\alpha) = \frac{\|n' \times n\|}{n' \cdot n} \frac{n' \cdot t}{|n' \cdot t|} \tag{6.24}$$

Where *n'* is computed as:

$$n' = \frac{X_3 - X_S + X_4 - X_S}{\|X_3 - X_S + X_4 - X_S\|} \text{ if } X_3 - X_4 \neq 0 \tag{6.25}$$

$$n' = \frac{X_4 - X_S}{\|X_4 - X_S\|} \begin{bmatrix} 0 & 1 \\ -1 & 0 \end{bmatrix} \text{ if } X_3 - X_4 = 0 \tag{6.26}$$

The contact element residual force vector then is modified as:

$$G_s = N_s P_{Ns} + \hat{T}_s P^e_{N\_cor} + \hat{T}_s T_{Ts} \tag{6.27}$$

The residual force that comes from the contact contribution is summed up with other nodal forces at each node to compute nodal erroneous residual force after each iteration of the Newton-Raphson method. The solution algorithm of the Newton-Raphson method also requires the linearization of the contact residual which is described in Section 6.2.4.

6.2.3 Constitutive relation at the contact interface

The hard contact is employed for normal contact, and the kinematic constraint ($g_{Ns} = 0$) is enforced by the penalty method. The penalty factor ε_N is selected to be very large such that the contact penetration is relatively small (in the case of this study, the contact penetration stands for the deviation of the cross-section centroid of a prestressing tendon from the cross-section centroid of sheathing). In the case of PT members with perfectly bonded tendons, slave nodes are stick to tendon sheathing segments. The same penalty factor is applied to both the normal and tangential directions at the contact interface. In the case of PT members with perfectly unbonded tendons, the contact pressure is only presented in the direction of normal contact. Thus, the tangential part of the contact element residual and the corresponding tangential stiffness is omitted. In the case of partially bonding, which is the most realistic situation, the contact tangential force is determined from the integration of a chosen frictional law, since the constitutive behavior in the tangential direction is very complex, and the goal of this study zeros in on developing the fundamental framework. The simple Coulomb's friction law is employed which assumes that the maximum tangential contact pressure is proportional to the normal contact pressure as:

$$f = \mu p_N \tag{6.28}$$

where μ is the frictional coefficient at the interface between the prestressing tendon and corresponding sheathing.

Accordingly, the slip determining function is defined as:

$$\text{Slip if } \hat{f}_s(t_T) = \|t_T\| - \mu p_{Nn} > 0 \tag{6.29}$$

where t_T is the trial tangential contact pressure at current state.

The simple Coulomb model yields a non-smooth tangential pressure sliding relation describing stick/sliding behavior which leads to mathematical difficulties (Figure 6.6). To overcome this issue, the tangential motion is split into an elastic (stick) and plastic (slip) part. An updated algorithm is used to update the tangential contact stress t_{Tn+1} at time step *n*+1. The details of the algorithm are described as follows:

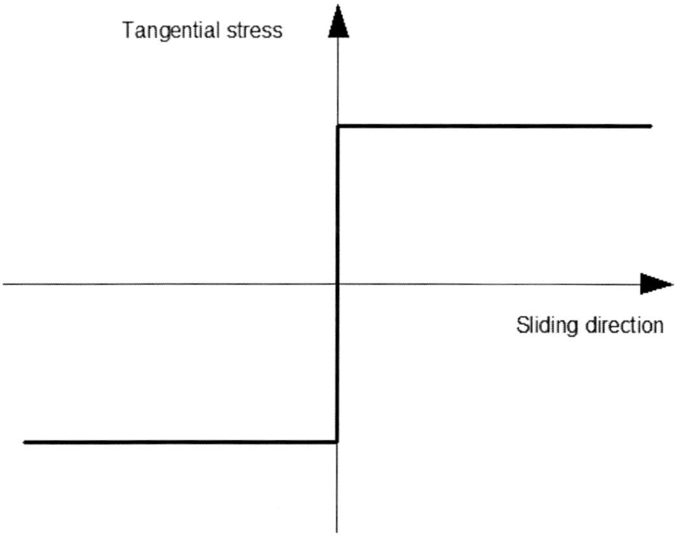

Figure 6.6 Coulomb's frictional model

In the general frame work of the Newton-Raphson method, the increment of sliding within a time step Δt_{n+1} (time step Δt_{n+1} only represents one iteration process within a load increment. It has nothing to do with the time itself as no time-dependent effects are considered in this study) is given by:

$$\Delta g_{Tn+1} = l\left(\bar{\xi}_{n+1} - \bar{\xi}_n\right)t_{n+1} \tag{6.30}$$

The stick/slip condition is unknown at this stage since this total increment must be subdivided into an elastic (stick) and plastic (slip) part. For a typical simulation of PT unbonded beams, the initial slip g_{T0}^{s} at each tendon node is zero.

$$g_{Tn+1}^{e} = g_{Tn+1} - g_{Tn}^{s} \tag{6.31}$$

where g_{Tn+1}^{e} represents the stick (elastic) part of the sliding at time *n+1*. g_{Tn}^{s}represents the plastic (slip) part of the sliding at time n.

By assuming that only stick behavior is taking place, a trial tangential contact stress is computed as

$$t_{tn+1}^{tr} = c_T\left(g_{Tn+1} - g_{Tn}^{s}\right) \tag{6.32}$$

where c_T is a very large elastic constant since the relative elastic sliding is usually very small (approximate stick behavior). In this study c_T is equal to normal contact penalty factor ε_N. This trial stress is inserted into Equation 6.29 to determine the slip condition as:

$$f_{sn+1}^{tr} = \|t_{Tn+1}^{tr}\| - \mu p_{Nn+1} \tag{6.33}$$

If $f_{sn+1}^{tr} \leq 0$, no slip takes place and the tangential stress at time step t_{n+1} is given by$t_{tn+1} = t_{tn+1}^{tr}$. If the slip condition is not fulfilled at the time incrementΔt_{n+1}, then the tangential stress must be projected onto the admissible region. There are a great variety of algorithms for solving plastic problems. In this case, the update of tangential stress and sliding is easily accomplished by using the implicit Euler scheme as:

$$g_{Tn+1}^{s} = g_{Tn}^{s} + \lambda n_{Tn+1} \tag{6.34}$$

where $n_T = \frac{t_T}{\|t_T\|}$.

The projected stress is defined as:

$$t_{Tn+1} = t_{Tn+1}^{tr} - \lambda c_T n_{Tn+1} \tag{6.35}$$

where $n_{Tn+1} = n_{Tn+1}^{tr}$.

In the case where Coulomb's law is employed for the tangential contact constitutive relation, λ is explicitly computed as:

$$\lambda = \frac{\|t_{tn+1}^{tr}\| - \mu p_{Nn+1}}{c_T} \tag{6.36}$$

Inserting Equation 6.36 into Equations 6.34 and 6.35, the updated plastic slip part of the total sliding and the tangential stress are computed as:

$$g_{Tn+1}^{s} = g_{Tn}^{s} + \frac{\left(\|t_{tn+1}^{tr}\| - \mu p_{Nn+1}\right) n_{Tn+1}^{tr}}{c_T} \tag{6.37}$$

$$t_{Tn+1} = \mu p_{Nn+1} n_{Tn+1}^{tr}$$

The algorithm described above requires storing the solutions of plastic slip of slave nodes at each time incrementΔt_{n+1}.

6.2.4 Solution algorithm

The most commonly used solution algorithm for nonlinear problems is the Newton-Raphson scheme. This method requires a linearization of contact residual. In this study, the tangential stiffness matrix of contact element is numerically evaluated by the complex variable differentiation method (Squire and Trapp, 1998). This method is not subject to subtraction error, from which conventional finite difference method usually suffers. Since the solution at each load step is either stick or slip, it is hard to determine the tangential stiffness in advance. Therefore, it is better to assume stick behavior of all contact elements at the beginning iteration of every load step. The algorithm of general computational procedures is described in Figure 6.7.

1. Assemble global K_{Ts}^c and K_{Ns}^c.
2. Add K_{Ts}^c and K_{Ns}^c to the global total tangential stiffness matrix K_T.
3. Compute displacement increment U at load step n.
4. Compute total tangential sliding distance G_{Ts} and total normal deviation distance G_{Ns} of each contact element at load step n.
5. Compute corresponding tangential and normal contact residual forces at load step n and assemble them to global total residual force vector.
6. Check if convergence criteria are met.

Figure 6.7 Flowchart of computational procedures

6.3 Embedding Element

The embedding element is essentially a two-node elastic Euler-Bernoulli beam element with very large axial and bending stiffness. The elastic constant *EA* and *EI* are sufficiently large such that the embedding element is treated as a rigid beam connecting the tendon sheathing and the reference axis of the PT frame member. The formulation of this embedding element follows the formulation of the traditional elastic Euler-

Bernoulli beam elements which is found in many literatures (e.g., Zienkiewicz and Taylor, 2005; Fish and Belytschko, 2007). Therefore, simple descriptions are given below.

The configuration of the two-node elastic Euler-Bernoulli beam element is shown in Figure 6.8. It contains two nodes and of each have two translational degrees-of-freedom and one rotational degree-of-freedom.

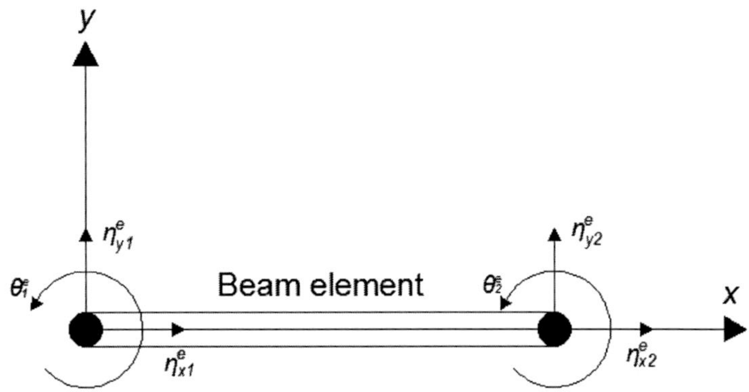

Figure 6.8 Two-node Euler-Bernoulli beam element

The linear interpolation is used along the axial axis of the beam and the 3rd order Hermite interpolation is used in interpolating transverse displacement. The element displacement vector is defined as:

$$\eta^e = [\eta_{x1} \quad \eta_{y1} \quad \eta_{\theta 1} \quad \eta_{x2} \quad \eta_{y2} \quad \eta_{\theta 2}]^T \tag{6.38}$$

where e denotes the element label.

The differential equation (strong form) governing the axial deformation is described as:

$$\frac{d}{dx}\left(AE\frac{d\eta_x}{dx}\right) + b = 0, 0 < x < l \tag{6.39}$$

where A is the cross-section area, E is the elastic constant and b is the body force.

Boundary conditions are specified as:

$$\sigma(0) = \left(E\frac{d\eta_x}{dx}\right)_{x=0} = \frac{P(0)}{A(0)} = -\bar{t} \tag{6.40}$$
$$\eta_x(l) - \bar{\eta}_x$$

where σ represents the normal stress. $\bar{t}$ represents the traction at boundary. $\bar{\eta}$ represents prescribed displacement vector at boundary.

The weak form is derived from the strong form as:

$$\int_0^l \frac{dw}{dx} EA \frac{d\eta_x}{dx} dx = (wA\bar{t})_{x=0} + \int_0^l wbdx \tag{6.41}$$

where w represents an arbitrary function (weighting function).

The first order interpolation used for axial deformation is defined as:

$$N_{\text{axial}}^e = \begin{bmatrix} \frac{x_2^e - x}{l^e} & 0 & 0 & \frac{x - x_1^e}{l^e} & 0 & 0 \end{bmatrix} \tag{6.42}$$

where x_1^e and x_2^e are the global x-coordinates of the first and second nodes of the beam element. x represents the global x-coordinate of an arbitrary point on the beam element.

The first derivative of the interpolation function is defined as:

$$B_{\text{axial}}^e = \frac{dN_{\text{axial}}^e}{dx} = \frac{1}{l^e}[-1 \quad 0 \quad 0 \quad 1 \quad 0 \quad 0] \tag{6.43}$$

Using the shape functions as weighting functions (Galerkin, 1915) while the cross-section area and elastic constant are the same through the beam, the element stiffness matrix is derived as:

$$k_{\text{axial}}^e = \int_0^{l^e} EAB^{e^T}B^e dx = \frac{EA}{l^e}\begin{bmatrix} 1 & 0 & 0 & -1 & 0 & 0 \\ 0 & 0 & 0 & 0 & 0 & 0 \\ 0 & 0 & 0 & 0 & 0 & 0 \\ -1 & 0 & 0 & 1 & 0 & 0 \\ 0 & 0 & 0 & 0 & 0 & 0 \\ 0 & 0 & 0 & 0 & 0 & 0 \end{bmatrix} \tag{6.44}$$

For bending part, the governing differential equation is as follows:

$$EI\frac{d^4\eta_y}{dx^4} - p = 0 \tag{6.45}$$

where P is a uniform load on the beam. I is the second moment of area of the cross section.

The boundary conditions are defined as:

$$
\begin{gathered}
\eta_y = \bar{\eta}_y \text{ on } \Gamma_\eta \\
mn = EI\frac{d^2\eta_y}{dx^2}n = \bar{m} \text{ on } \Gamma_m \eta_y = \bar{\eta}_y \text{ on } \Gamma_\eta \\
\frac{d\eta_y}{dx} = -\bar{\theta} \text{ on } \Gamma_\theta \\
sn = -EI\frac{d^3\eta_y}{dx^3}n = \bar{s} \text{ on } \Gamma_s
\end{gathered}
\tag{6.46}
$$

where n is the normal, m is the moment and s is the shear on the boundary. $\bar{\theta}$ represents the prescribed rotation at boundary.

The weak form is derived as:

$$
\int_0^l \frac{d^2w}{dx^2}EI\frac{d^2\eta_y}{dx^2}dx = \int_0^l wpdx + \left(\frac{dw}{dx}\bar{m}\right)\Big|_{\Gamma_m} + (w\bar{s})|_{\Gamma_s} \tag{6.47}
$$

The 3rd order Hermite interpolation functions are defined as:

$$
N^e_{\text{bending}} = \begin{bmatrix} 0 \\ \frac{1}{4}(1-\xi)^2(2+\xi) \\ \frac{l^e}{8}(1-\xi)^2(1+\xi) \\ 0 \\ \frac{1}{4}(1+\xi)^2(2-\xi) \\ \frac{l^e}{8}(1+\xi)^2(\xi-1) \end{bmatrix}^T \tag{6.48}
$$

where ξ is the local coordinate of the beam element ($-1 \le \xi \le 1$).

The second derivative of the interpolation functions is derived as:

$$B_{\text{bending}}^{e} = \frac{d^2 N_{\text{bending}}^{e}}{dx^2} = \frac{1}{l^e}\begin{bmatrix} 0 & \frac{6\xi}{l^e} & 3\xi - 1 & 0 & -\frac{6\xi}{l^e} & 3\xi + 1 \end{bmatrix} \tag{6.49}$$

Based on the Galerkin method (Galerkin, 1915) and constant elastic modulus and cross section area through the element, the element stiffness matrix is explicitly expressed as:

$$k_{\text{bending}}^{e} = \int_0^{l^e} EI{B^e}^T B^e dx = \frac{EI}{{l^e}^3}\begin{bmatrix} 0 & 0 & 0 & 0 & 0 & 0 \\ 0 & 12 & 6l^e & 0 & -12 & 6l^e \\ 0 & 6l^e & 4{l^e}^2 & 0 & -6l^e & 2{l^e}^2 \\ 0 & 0 & 0 & 0 & 0 & 0 \\ 0 & -12 & 6l^e & 0 & 12 & -6l^e \\ 0 & -6l^e & 2{l^e}^2 & 0 & -6l^e & 4{l^e}^2 \end{bmatrix} \tag{6.50}$$

The embedding element stiffness matrix is the sum of the axial stiffness and bending stiffness as:

$$k_{\text{embedding}}^{e} = k_{\text{axial}}^{e} + k_{\text{bending}}^{e} \tag{6.51}$$

Since the element behaves linearly, the tangential stiffness matrix is the same as the stiffness matrix derived above.

6.4 Anchorage Element

6.4.1 Treatment for tendon jacking

Prestressing forces are enforced by the procedure of tendon jacking in practice. Several methods have been developed and successfully applied to numerical models for simulating jacking procedures by means of FEM (e.g., Nikolic and Mihanovic, 1996; Stavroulaki et al., 1997). A common treatment of tendon jacking is to connect prestressing tendons to the anchorages before applying prestressing force. An initial strain or stress field is directly applied to the elements of tendons afterwards. The initial strain or stress field can be approached directly or indirectly. The latter manipulates a temperature field around the tendon elements, usually by cooling down the tendon to achieve the desired prestressing level (Stavroulaki et al., 1997). In the case of simulating PT tendon jacking, the initial stress/strain field approach has a drawback that neglects the prestress loss during the tendon jacking procedure, as the initial strain generated around the tendon is uniformly distributed. To overcome this issue, another adopted approach is to simulate the mechanical behavior of tendon jacking in real practice (Vecchio et al., 2006). The simulation of tendon jacking and anchorage process introduces the so called boundary nonlinearity. In such a case, not only is a contact formulation required to imitate the real mechanical behavior but a special treatment also needs to be developed. In this study, a special procedure is introduced to deal with the boundary nonlinearity at the anchorage zone. The treatment is enforced by a special anchorage element along with some modifications to the Newton-Raphson method.

6.4.2 General procedures

The anchorage element is developed to serve as an interface element between a tendon element and an embedding element. This element is only effective after tendon jacking and the completion of anchorage wedge setting. The anchorage element comes into play as a very rigid spring between the anchor point and tendon end as shown in Figure 6.9.

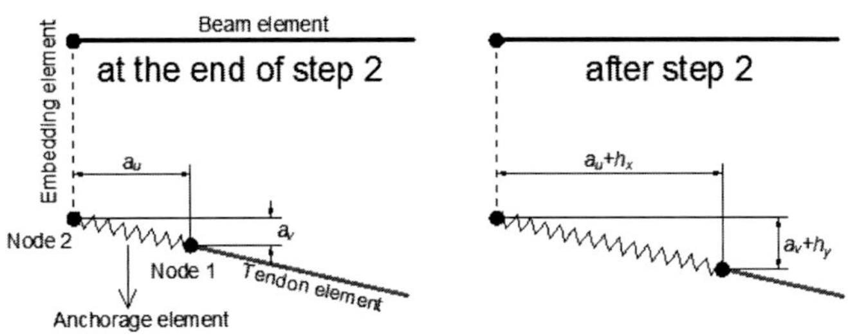

Figure 6.9 Anchorage element

Only the translational degrees-of-freedom of Node1 and Node2 are constrained to each other by the penalty method. Good accuracy is obtained if the penalty factor is sufficiently large to eliminate any relative motion between the tendon end and anchor point. In order to fully simulate the prestressing process, a typical finite element simulation is subdivided into four steps as follows: 1) pulling out tendon; 2) anchorage wedge loss; 3) anchoring tendon; 4) service loading. At steps one and two, the tendon end node (for example, Node 1 in Figure 6.8) has an essential boundary condition, and a corresponding prescribed displacement load is applied to simulate the distance of the tendon pulling out and subsequent anchorage setting. At each iteration process, the reaction forces are computed at the tendon end node and applied to the embedding element node connecting to the tendon (for example, Node2 in Figure 6.8). This treatment ensures that the jacking reactions are applied to the PT member itself. The schematic view of this process is shown in Figure 6.10.

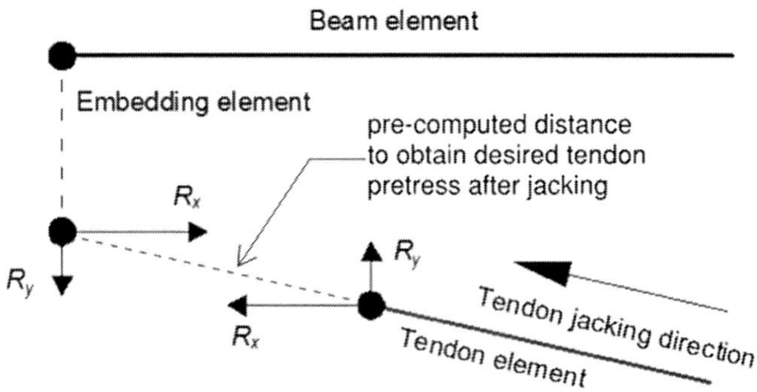

Figure 6.10 Tendon jacking reactions transferred to PT members

Once equilibrium is achieved at the end of step two, a tendon end node and one embedding element node serve as anchor point that are supposed to be very close to each other. This is achieved by pre-calculating the desired tendon pulling force and anchorage slip. The pre-calculated distances are used to determine the initial coordinates of the tendon end node such that after tendon pulling out and drawback, tendon end nodes are close to their corresponding anchor points. There will always be a small distance between the tendon end node and corresponding anchor point due to the difficulty of accurately determining the initial position of tendon end nodes. The x-component of this distance in global coordinates is depicted as a_u and y-component depicted as a_v in Figure 6.8. At step three, anchorage elements are added to the system. In the case of one end jacking, only one anchorage element is added to the formulation, otherwise two anchorage elements are inserted in the case of two end jacking. In addition, the essential boundary condition at tendon end and the corresponding reaction forces acting on anchor points are removed. Due to the stiffness of added anchorage elements being sufficiently large (typically one order of magnitude larger than the typical stiffness of the structural system), negligible relative motions are generated once the equilibrium is reached at the end of step three. At present, for any displacement increment occurs within and after step three, the added part of displacements are denoted as h_x and h_y . Therefore, the entire displacements of tendon end nodes are sub-divided into two parts as displacements generated at the end of step two, and the displacements generated after step two. The linear stiffness of the anchorage element is only related to the displacements generated after step two, which makes the constitutive equation of the anchorage element nonlinear. A linearization of element residual forces is required to obtain the tangential stiffness matrix in the solution process. The general procedures of the special treatment are described in Figure 6.11.

1. Estimate initial tendon end node positions such that the end nodes are roughly overlapped with corresponding anchor points after the tendon jacking and wedge setting loss.
2. At analysis step one and two, set tendon end nodes as essential boundary conditions and apply displacement of tendon pulling force and anchorage slip; apply reactions of tendon end nodes to corresponding anchor points as external loads at beginning of each iteration process.
3. At analysis step 3, change boundary condition of tendon end nodes to natural boundary type; add anchorage element between tendon end nodes and corresponding anchor points. The process of applying reactions at anchor points is removed.
4. At analysis step 4, apply service load.

Figure 6.11 General procedure of tendon jacking and anchoring

6.4.3 Anchorage element formulation

The anchorage element contains two nodes. The first node is a tendon end node while the second node is an embedding element node. The displacement vector of one anchorage element is written as:

$$\eta^e = [\eta_{x1} \quad \eta_{y1} \quad \eta_{\theta 1} \quad \eta_{x2} \quad \eta_{y2} \quad \eta_{\theta 2}]^T \tag{6.52}$$

where e denotes the element label.

Let us denote the relative motion between tendon end node and corresponding embedding element node as h_x and h_y in the direction of global x and y directions, respectively. At the end of analysis step two, the displacement vector of the anchorage element is defined as:

$$d_a^e = [\eta_{x1} \quad \eta_{y1} \quad \eta_{\theta 1} \quad \eta_{x2} \quad \eta_{y2} \quad \eta_{\theta 2}]^T \tag{6.53}$$

Accordingly,

$$\begin{aligned} h_x^e &= (\eta^e - d_a^e)^T N_1 \\ h_y^e &= (\eta^e - d_a^e)^T N_2 \end{aligned} \tag{6.54}$$

$$\begin{aligned} N_1 &= [1 \quad 0 \quad -1 \quad 0 \quad 0]^T \\ N_2 &= [0 \quad 1 \quad 0 \quad -1 \quad 0]^T \end{aligned} \tag{6.55}$$

If the penalty form is used, the virtual work contribution from one anchorage element is computed as:

$$\delta\Pi = \varepsilon_{ac} h_x^e N_1^T \eta^e + \varepsilon_{ac} h_y^e N_2^T \eta^e \tag{6.56}$$

where ε_{ac} is the penalty factor of anchorage element.

The residual force vector of one anchorage element is written as:

$$G_e = \left(\varepsilon_{ac} h_x^e N_1^T + \varepsilon_{ac} h_y^e N_2^T\right)^T \tag{6.57}$$

Substituting Equation 6.54 into Equation 6.56 yields:

$$G_e = [\varepsilon_{ac}(\eta^e - d_a^e)^T N_1 N_1^T + \varepsilon_{ac}(\eta^e - d_a^e)^T N_2 N_2^T]^T \tag{6.58}$$

Since the constitutive relation of the anchorage element is nonlinear, a linearization of Equation 6.58 is required to incorporate with the Newton-Raphson method. By linearization of Equation 6.58, the tangential stiffness matrix of one anchorage element is derived as:

$$k_{ac}^e = \varepsilon_{ac} N_1 N_1^T + \varepsilon_{ac} N_2 N_2^T = \varepsilon_{ac} \begin{bmatrix} 1 & 0 & -1 & 0 & 0 \\ 0 & 1 & 0 & -1 & 0 \\ -1 & 0 & 1 & 0 & 0 \\ 0 & -1 & 0 & 1 & 0 \\ 0 & 0 & 0 & 0 & 0 \end{bmatrix} \tag{6.59}$$

6.5 Nonlinear Beam and Truss Elements

6.5.1 Nonlinear RC beam elements

In this section, a simple but powerful Euler-Bernoulli type beam element is introduced. It has two-nodes with three degrees-of-freedom per node. The proposed element is capable of considering material nonlinearity introduced by concrete cracking and reinforcement yielding. The tension stiffening effect is considered by modifying either the uniaxial stress-strain relation of concrete or mild steel. This effect, however, plays an insignificant role in numerical simulation of PT structures mostly due to the extremely low reinforcement ratio of typical PT structures. It also allows arbitrary shape of beam sections with arbitrary configurations of reinforcement, which however needs to be symmetric in the out of plane direction due to the nature of a two-dimensional analysis. A fiber section model is adopted for the section integration and the numerical integration is performed along the longitudinal axis of the beam. It is computationally efficient compared to other types of elements while still meeting the practical engineering requirements of accuracy. The behavior of the element is modeled based on proper simplification of beam kinematics and actual material laws. The limitation of the beam element is as follows. 1) The concrete cracking is smeared throughout an element, thus no individual crack will be tracked during a simulation; 2) Perfect bonding is assumed between the concrete and mild steel; 3) Strain predictions of mild steel might deviate from the actual case at the ultimate stage due to this assumption. This strain deviation, however, has a minor effect on the simulation as the intent of this study is to focus on PT structures which are usually very lowly reinforced.

6.5.2 Element formulation

The nonlinear RC beam element introduced in this study is extended from a conventional Euler-Bernoulli elastic beam element which has two nodes with six degrees-of-freedom in a 2-D plane as shown in Figure 6.12.

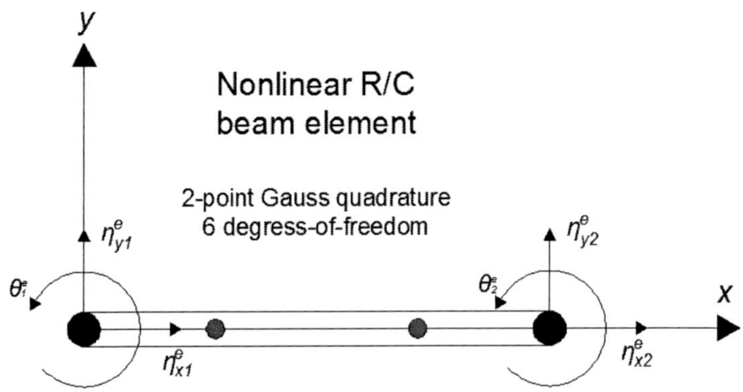

Figure 6.12 Two-node beam element

The transverse displacement is interpolated by the 3rd order Hermite polynomial, and a linear interpolation is employed for the axial deformation. It is noted that both the axial deformation and bending deformation depend on the axial strain interpolation. The axial strain interpolated from axial and bending deformation, therefore, should be in the same order. However, the linear interpolation in the axial direction results in a constant axial strain, whereas the axial strain interpolated from the 3rd order Hermite polynomial may not be the same. The effect of inconsistent interpolations of axial strain will be observed by a simulation with a coarse mesh but it will converge to a formulation with consistent axial strain when the number of elements increases. Considering the contact formulation introduced in the study, which approximates the tendon profile into several piecewise linear segments, the tendon geometry is recovered when the number of beam elements increases in a simulation as shown in Figure 6.13. Therefore, the combination of a linear and the 3rd order Hermite polynomial interpolations provide enough accuracy with a refined mesh from an engineering viewpoint.

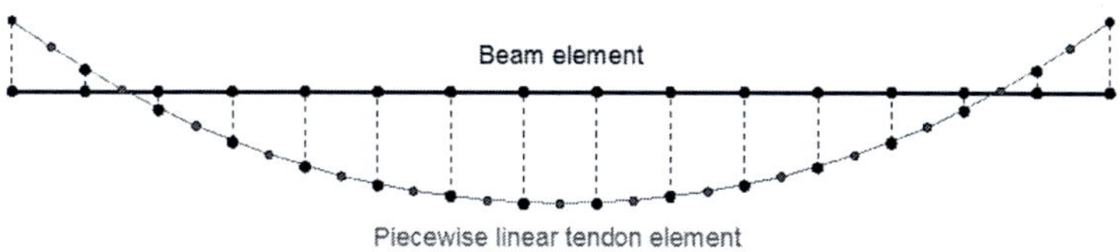

Figure 6.13 Piecewise linear tendon elements

The displacement vector of the nonlinear RC beam element is defined as

$$\eta^e = [\eta_{x1} \quad \eta_{y1} \quad \eta_{\theta 1} \quad \eta_{x2} \quad \eta_{y2} \quad \eta_{\theta 2}]^T \tag{6.60}$$

where e is the element label.

The weak forms for the axial and bending deformation are revoked as described in embedding element formulation. For axial deformation,

$$\int_0^l \frac{dw}{dx} EA \frac{d\eta_x}{dx} dx = (wA\bar{t})_{x=0} + \int_0^l wbdx \tag{6.61}$$

The first order interpolation used for axial deformation is defined as:

$$N_{\text{axial}}^e = [N_1 \quad N_4] = \begin{bmatrix} \frac{x_2^e - x}{l^e} & \frac{x - x_1^e}{l^e} \end{bmatrix} \tag{6.62}$$

For bending deformation,

$$\int_0^l \frac{d^2 w}{dx^2} EI \frac{d^2 \eta_y}{dx^2} dx = \int_0^l wpdx + \left(\frac{dw}{dx}\bar{m}\right)\Big|_{\Gamma_m} + (w\bar{s})|_{\Gamma_s} \tag{6.63}$$

The 3rd order Hermite interpolation functions are defined as:

$$N_{\text{bending}}^e = \begin{bmatrix} N_2 \\ N_3 \\ N_5 \\ N_6 \end{bmatrix}^T = \begin{bmatrix} \frac{1}{4}(1-\xi)^2(2+\xi) \\ \frac{l^e}{8}(1-\xi)^2(1+\xi) \\ \frac{1}{4}(1+\xi)^2(2-\xi) \\ \frac{l^e}{8}(1+\xi)^2(\xi-1) \end{bmatrix}^T \tag{6.64}$$

Consequently, the deferential operator matrix B is defined as:

$$B = \begin{bmatrix} \frac{\partial N_1}{\partial x} & 0 & 0 & \frac{\partial N_4}{\partial x} & 0 & 0 \\ 0 & \frac{\partial^2 N_2}{\partial x^2} & \frac{\partial^2 N_3}{\partial x} & 0 & \frac{\partial^2 N_5}{\partial x^2} & \frac{\partial \ N_6}{\partial x^2} \end{bmatrix} \tag{6.65}$$

The interpolation matrix N is defined as:

$$N = \begin{bmatrix} N_1 & 0 & 0 & N_4 & 0 & 0 \\ 0 & N_2 & N_3 & 0 & N_5 & N_6 \end{bmatrix} \tag{6.66}$$

The resultant force vector in a section is defined as:

$$P = \begin{Bmatrix} P_a \\ M_y \end{Bmatrix} \tag{6.67}$$

where P_a stands for the resultant axial force and M_y stands for the resultant moment in a section.

The body force vector as is defined as:

$$q = \begin{Bmatrix} q_x \\ q_z \end{Bmatrix} \tag{6.68}$$

where q_x represents the body force in the longitudinal direction. q_z represents the body force in the in-plane transverse direction.

Considering Equations 6.61 and 6.63, the finite element irreducible form for inelastic problems is derived and the element residual force vector is deduced from the virtual work contribution of beam element

$$\delta \Pi = \delta \eta^T \int_{l^e} (B^T P - N^T q)\, dx \tag{6.69}$$

The element stiffness matrix is determined according to the principle of virtual work as:

$$\bar{k}^e_{beam} = \int_{l^e} B^T D B dx \tag{6.70}$$

where D is expressed as:

$$D = \int_A \begin{bmatrix} E & zE \\ zE & z^2 E \end{bmatrix} \tag{6.71}$$

where A is the area of the section. E is the elastic modulus (reduced elastic modulus if material goes into inelastic range) obtained from local constitutive relations at each computation point in the section and z stands for the distance between the integration point and the reference axis of the beam (the sign of z depends on which side of reference axis the integration point is on according to the definition).

A linearization of the element residual force vector yields the tangent stiffness matrix as:

$$k^e_{beam} = \int_{l^e} B^T D_T B dx \tag{6.72}$$

where D_T is expressed as:

$$D_T = \int_A \begin{bmatrix} E_T & zE_T \\ zE_T & z^2 E_T \end{bmatrix} \tag{6.73}$$

where E_T is the material tangent modulus obtained from the local constitutive relations at each computation point in the section.

The matrices D and D_T are not diagonal if the material is inelastic or the reference axis of the beam is not on the center of gravity of the section. It is noted that if the reference axis refers to the centroidal axis and the material is elastic, matrix D becomes diagonal.

In order to compute D and D_T matrices, a fiber model is introduced to perform the discrete integration on the section. A beam section is divided into a number of sub-fiber sections. In this study, rectangular shaped fibers are used to approximate the beam section as shown in Figure 6.14.

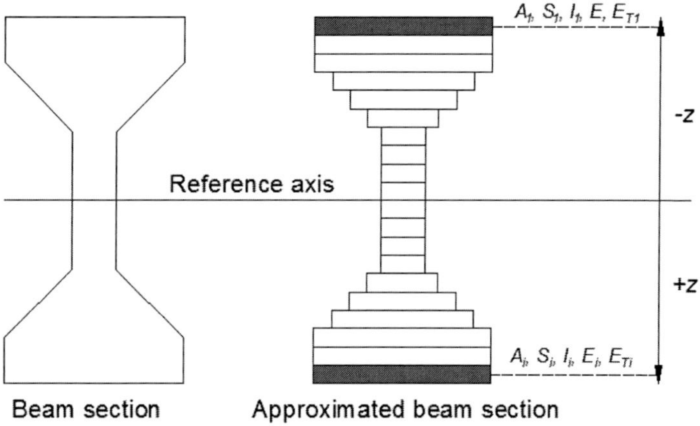

Figure 6.14 Approximated fiber-based section

As shown in the figure, the discretely integrated area, the first moment of area, second moment of area (moment of inertia), elastic modulus and tangent modulus converge to the exact integration of the original section as the number of fibers n increases. Therefore, the integration of matrices D and D_T are approximated as:

$$D = \int_A \begin{bmatrix} E & zE \\ zE & z^2E \end{bmatrix} \approx \sum_{i=1}^{n} \begin{bmatrix} A^iE^i & S^iE^i \\ S^iE^i & I^iE^i \end{bmatrix} \tag{6.74}$$

where D_T is expressed as:

$$D_T = \int_A \begin{bmatrix} E_T & zE_T \\ zE_T & z^2 E_T \end{bmatrix} \approx \sum_{i=1}^{n} \begin{bmatrix} A^i E_T^i & S^i E_T^i \\ S^i E_T^i & I^i E_T^i \end{bmatrix} \tag{6.75}$$

where S is the first moment of area. I is the second moment of area. n is the number of fibers in a section.

If there is reinforcement in the section, the mild steel bars are considered as separate individual fibers according to their positions. As a result, matrices D and D_T consist of contributions from both concrete and reinforcement as:

$$\begin{aligned} D &= D^c + D^s \\ D_T &= D_T^c + D_T^s \end{aligned} \tag{6.76}$$

where D^c and D^s are D matrices constructed from concrete and steel reinforcement, respectively.

The computation of section integration follows the same manner described above for both concrete and steel. The integration along the longitudinal direction of the beam is performed by the Gauss quadrature. Some researchers reported that a six points Gauss integration proved to be the optimum between computation time and accuracy of results for this type of a beam element (Pőttler and Swoboda, 1987). Considering a single element length is limited to a small value, but with a large number of elements assembled in the system, a two point Gauss integration yields sufficient accuracy. The two point Gauss quadrature was selected and implemented in the program as Table 6.1 shows the position and weighting of each integration point.

Table 6.1 Gauss-Legendre rule

<table>
<tr><th>Number of points</th><th>Points</th><th>Weight</th><th>Degree</th></tr>
<tr><td>2</td><td>±0.57735027</td><td>1</td><td>3</td></tr>
<tr><td rowspan="2">3</td><td>0</td><td>0.88888889</td><td rowspan="2">5</td></tr>
<tr><td>±0.77459667</td><td>0.55555555</td></tr>
<tr><td rowspan="2">4</td><td>±0.33998104</td><td>0.65214515</td><td rowspan="2">7</td></tr>
<tr><td>±0.86113631</td><td>0.34785485</td></tr>
<tr><td rowspan="3">5</td><td>0</td><td>0.56888889</td><td rowspan="3">9</td></tr>
<tr><td>±0.53846931</td><td>0.47862867</td></tr>
<tr><td>±0.90617985</td><td>0.23692689</td></tr>
<tr><td rowspan="3">6</td><td>±0.23861918</td><td>0.46791393</td><td rowspan="3">11</td></tr>
<tr><td>±0.66120939</td><td>0.36076157</td></tr>
<tr><td>±0.93246951</td><td>0.17132449</td></tr>
</table>

At this point, the element residual force vector and tangent stiffness matrix for one element are derived, and the Newton-Raphson type method is applied to obtain the solution.

6.5.3 Constitutive relations for concrete and steel materials

The aforementioned formulation requires the computation of elastic modulus, reduced elastic modulus and tangent modulus for each computation point in the section. These material parameters are obtained from uniaxial constitutive relations of both concrete and steel. Once the material goes into inelastic state, the reduced elastic modulus is defined as the current stress divided by the current strain. The concrete uniaxial stress-strain relation in compression is determined by an empiric equation proposed by Carreria and Chu (1985). As for tension, concrete is considered elastic until tensile strength, f_{cr} is reached. After cracking, stress drops to a very small value in order to maintain numerical stability. The tension stiffening effect is considered by modifying the descending portion of stress-strain relations after cracking. With typical non-prestressed RC structures, the tension stiffening effect is quite noticeable due to the large reinforcement ratio, in which case, the stress might linearly reduce to the residual stress after cracking at a strain of about ten times the strain at cracking. On the other hand, typical PT structures have very low reinforcing ratio in comparison with conventional RC structures. The tension stiffening effect is so miniscule that one can assume the stress in tension drops to the residual stress at a strain of about two times that at cracking. This assumption is consistent with simulations of PT slabs and beams carried out via general purpose finite element packages in Chapter 5. A typical concrete uniaxial stress-strain relation is shown in Figure 6.15, and it is the basis of concrete material behavior in this research.

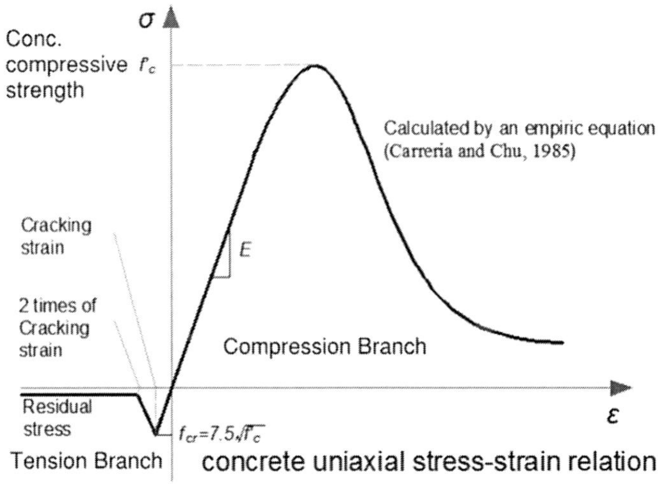

Figure 6.15 Uniaxial stress-strain relation assumed for concrete material

As for mild steel bar, it is assumed that its behavior follows a bilinear relation as shown in Figure 6.16. Steel material remains elastic until yielding stress, f_y is reached. This assumption is fairly accurate for the current study as a yielding plateau is quite often observed in coupon tests of mild steel bars. Strain hardening is modeled since the implementation method chosen only requires a discrete stress-strain relation. Once the stress-strain relations are determined for both concrete and steel, a piecewise linear interpolation is made to the stress-strain relations, and the corresponding elastic modulus (or reduced elastic modulus) and tangent modulus.

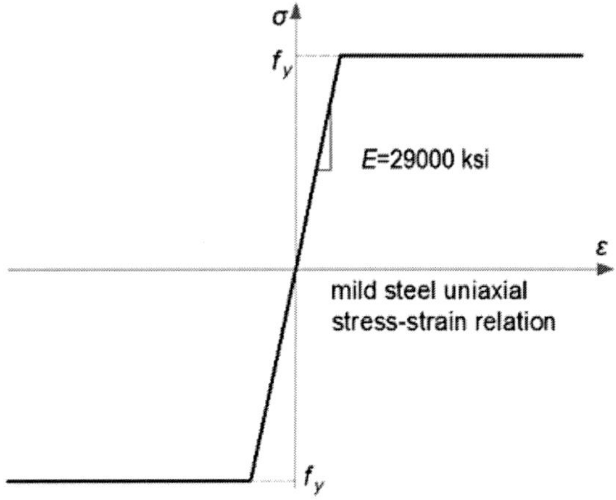

Figure 6.16 Uniaxial stress-strain relation assumed for steel material

6.5.4 Solution process

Once the uniaxial stress-strain relation of concrete and steel is determined, the elastic modulus or reduced elastic modulus and tangent modulus are computed at each computation point in a section. In the Newton-Raphson solving process, the strain and curvature at each integration point are measured based on the current displacement vector. Following the assumption that plane sections remain plane for a Euler-Bernoulli type beam, the strain distribution in a section is linear as shown in Figure 6.17.

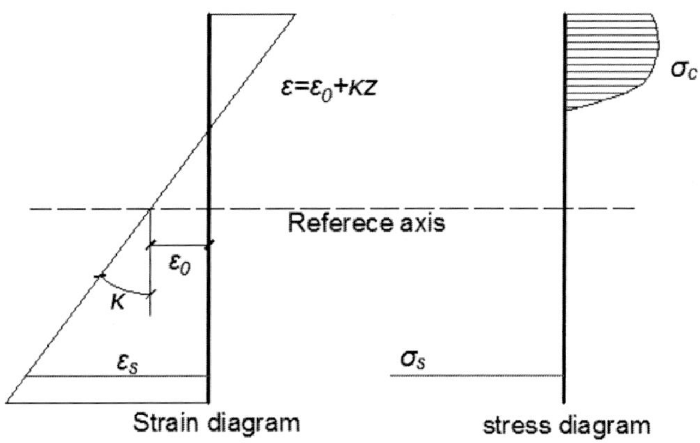

Figure 6.17 Stress and strain distribution in a typical beam section

The axial strain at each centroid of fiber sections is determined by:

$$\varepsilon = \varepsilon_0 + \kappa z \tag{6.77}$$

where ε_0 and κ are the axial strain and curvature measured at the reference axis of a section, respectively. z remains the same definition in Equation 6.71.

Once strains at all fiber locations are known, interpolations are made to compute the element residual force vector as well as the tangent stiffness matrix. The residual force vector and tangent stiffness matrix are used to obtain the next displacement increment vector.

6.5.5 Tendon element

The elements of prestressing tendons are conventional two-node truss elements limited to small strain and rotations. The formulation of the inelastic two-dimensional rod element is quite similar to the axial deformation that has been discussed in the nonlinear beam formulation. It is a two nodes element with four degrees-of-freedom as shown in Figure 6.18.

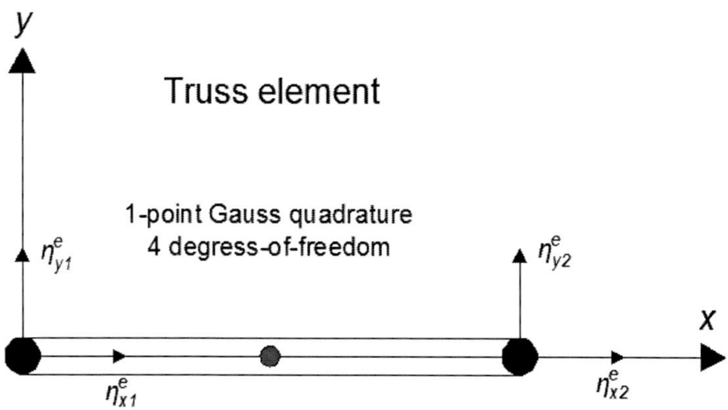

Figure 6.18 Two-node truss element

The displacement vector of the truss element is defined as:

$$\eta^e = [\eta_{x1} \quad \eta_{y1} \quad \eta_{x2} \quad \eta_{y2}]^T \tag{6.78}$$

where e is the element label.

The stiffness matrix at current displacement is computed by a one point Gauss quadrature or analytical integration as:

$$\bar{k}^e_{tendon} = \int_0^{l^e} EAB^{e^T}B^e dx = \frac{EA}{l^e}\begin{bmatrix} 1 & 0 & -1 & 0 \\ 0 & 0 & 0 & 0 \\ -1 & 0 & 1 & 0 \\ 0 & 0 & 0 & 0 \end{bmatrix} \quad (6.79)$$

where E is the elastic modulus or reduced elastic modulus obtained at the local constitutive relation. A is the area of the section of one truss element. l^e is the element length.

Similarly, the tangent stiffness matrix is derived as:

$$k^e_{tendon} = \int_0^{l^e} E_T AB^{e^T}B^e dx = \frac{E_T A}{l^e}\begin{bmatrix} 1 & 0 & -1 & 0 \\ 0 & 0 & 0 & 0 \\ -1 & 0 & 1 & 0 \\ 0 & 0 & 0 & 0 \end{bmatrix} \quad (6.80)$$

where E_T is the tangent modulus obtained at the local constitutive relation.

The proposed implementation method requires a uniaxial stress-strain relation defined for prestressing tendons. The stress-strain relation is either obtained from field tests or simply from empiric equations such as one developed by Devalapura and Tadros (1992) for Grade 270 seven-wire strands. A typical uniaxial stress-strain relation for Grade 270 seven-wire strands is shown in Figure 6.19.

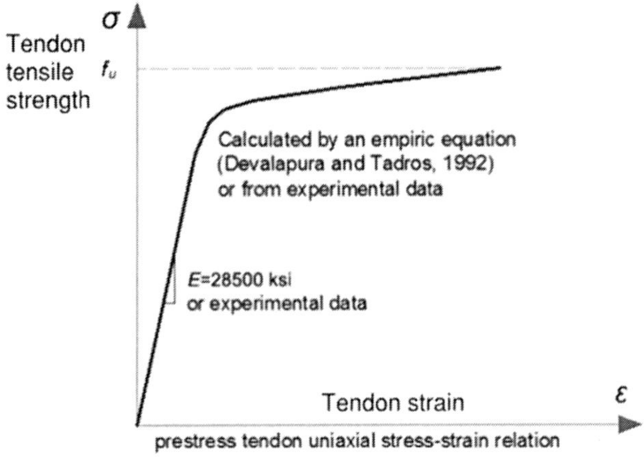

Figure 6.19 Typical uniaxial stress-strain relation of the prestressing tendon

6.6 Solution Method

6.6.1 Solving method for nonlinear equations

In solid mechanics, structural nonlinear finite element problems typically are categorized into three sources as material, geometrical and boundary nonlinearities. The most frequently applied solution scheme is the Newton-Raphson algorithm and its numerous variants. This method is based on a Taylor series expansion at a known state. If the force equilibrium of a finite element system is expressed as:

$$G(\eta, \lambda) = 0 \tag{6.81}$$

where η is the state variable as displacement vector and λ is the factor of load vector.

Then, the Taylor series expansion of Equation 6.81 yields:

$$G(\eta_k + \Delta\eta, \bar{\lambda}) = G(\eta_k, \bar{\lambda}) + DG(\eta_k, \bar{\lambda})\Delta\eta + r(\eta_k, \bar{\lambda}) \tag{6.82}$$

where $DG(\eta_k, \bar{\lambda})$ is the gradient of G evaluated at $\eta = \eta_k, \lambda = \bar{\lambda}$. $r(\eta_k, \bar{\lambda})$ is the residual part.

The linearization of G yields a matrix, which is the Jacobian or also known as tangential stiffness matrix denoted K_T. The residual vector $r(\eta_k, \bar{\lambda})$ of Equation 6.82 is neglected due to its higher order. By neglecting the high order term, a new equation is obtained as:

$$G(\eta_k, \bar{\lambda}) + DG(\eta_k, \bar{\lambda})\Delta\eta = 0 \tag{6.83}$$

Equation 6.83 becomes the basis of the following iterative algorithm for the solutions of nonlinear problems. The displacement increments, $\Delta\eta$, at state k is computed by solving a set of linear algebraic equations. Therefore, the solution of the nonlinear algebraic equation set is converted to solving a series of linear algebraic equation sets. The treatment of boundary conditions in solving the linear system is described in a later section. To assemble the global tangential stiffness matrix, all sub-tangential stiffness matrices are summed together (tangential stiffness matrices from anchorage, beam, contact, embedding, and tendon elements). For a variety of elements developed in this study, the corresponding analytically derived tangential stiffness matrices were already shown in previous sections. The quadratic convergence of the Newton-Raphson scheme is only valid if the initial value is near the solution point (Wriggers, 2008). Therefore, to apply a load in several steps by a load factor λ usually ensures that the solution is converged locally and efficiently at each loading level. The Newton-Raphson scheme is illustrated in Figure 6.20.

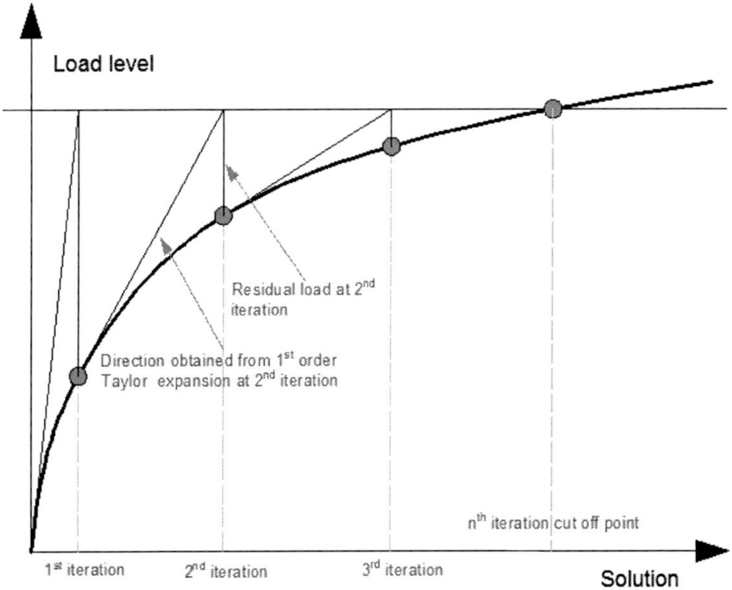

Figure 6.20 Illustration of Newton-Raphson scheme

6.6.2 Line search modification

The convergence rate of the Newton-Raphson scheme sometimes is very slow even if the load is applied in several steps. A diverged solution is more likely obtained if a very high nonlinearity presents in the system. To increase the robustness of the Newton-Raphson scheme, the line search technique is applied to the solving process. The rationale behind this method is that the direction of the displacement increment vector $\Delta\eta$ found by the Newton-Raphson scheme is often a good direction. The step size, however, is not optimal. It is cheaper to find the optimal step size within one step than to obtain a new direction by computing new tangential stiffness matrices. When a new displacement increment vector is obtained, the corresponding residual force vector is computed. Therefore, before proceeding to the next direction, the residual force vector is minimized by applying a factor γ to the current displacement incrementsη_k at state k. In this study, the l_2 norm of the residual force vector is considered as the minimization measure. This is expressed as:

$$\text{minimize} \left\|G\left(\eta_{lr} + \gamma\Delta\eta, \bar{\lambda}\right)\right\|_{l_2} \quad \gamma > 0 \qquad (0.84)$$

Once a measure of the residual force vector has been established, the line search is made with any of the methods for minimizing a function of a single parameter. The method of bisection is selected here for the current study. The implementation of this method is shown in Figure 6.21.

1. Set $tolerance = tol \quad 0 < tol < 1 \quad \eta = \Delta\eta$ (current displacement increment) $loop = 1$.
2. Compute residual force r_f according to η
3. Set $\eta_1 = 0 \quad \eta_2 = \eta \quad flag = 0$
4. Loop over until $flag = 1$

    $\eta_r = (1 + tol)\eta$
    $\eta_l = (1 - tol)\eta$
    compute residual r_1, r_2 and r_3 according to η_1, η and η_2
    compute corresponding l^2 norm of r_1, r_2 and r_3 as $norm_1$, $norm_2$ and $norm_3$
    if $norm_1 < norm_2 < norm_3$
        $\eta_2 = \eta$
        $\eta = 0.5(\eta_l + \eta_r)$
        compute residual r_2 according to η
    elseif $norm_1 > norm_2 > norm_3$
        if $loop = 1$
            increase η until the tangent at η is positive
            $\eta_2 = \eta$
        else
            $\eta_1 = \eta$
            $\eta = 0.5(\eta_l + \eta_r)$
            compute residual r_2 according to η
        end
    elseif $norm_1 > norm_2$ and $norm_2 < norm_3$
        $flag = 0$
    end
    $loop = loop + 1$

Figure 6.21 Flow chart of bisection method

6.6.3 Solution control and convergence criteria

The termination of the iterative procedure in solution process by the Newton-Raphson method is determined by convergence criteria. The selected criteria for the current study are based on the magnitude of the residual force vector as well as the magnitude of the displacement increments vector. The l_2 norm of residual force vector or displacement increment vector is computed and compared with the l_2 norm of the internal force vector or total displacement at each iteration process. The tolerance coefficient ε of error determines the precision level of the computed displacement vector before terminating the iterative process. It is shown in Equation 6.85.

$$\|r\|_{l_2} = \sqrt{\sum_{a=1}^{n} r_a^2} \leq \varepsilon \left\| f_{int\,l_2} \right\| \tag{6.85}$$

$$\|\Delta\eta\|_{l_2} = \sqrt{\sum_{a=1}^{n} (\Delta\eta_a)^2} \leq \varepsilon \|\eta\|_{l_2}$$

where a is the global degrees-of-freedom label and n is number of the degrees-of-freedom in the system. r is the residual force. $\Delta\eta$ is the displacement increment.

The convergence tolerance determines the computational work and accuracy. If it is too coarse, the solution might be quite inaccurate, while a too tight criterion results in unnecessary computational works. For reinforced concrete structural analysis, due to the nature of very high nonlinearity presented in concrete material, a good practice is $\varepsilon = 1\%$ from an engineering viewpoint.

6.6.4 Solution process of linear system equations

Once the global tangential stiffness matrix and external force vector are assembled, a linear algebraic equation system is constructed as:

$$K_T \Delta\eta = \Delta f \tag{6.86}$$

where Δf is the load increment vector.

The displacement increment vector is only computed after imposing appropriate essential boundary conditions (displacement boundary conditions). There are many approaches to inserting the displacement boundary conditions such as the partition method, penalty factor, etc. Here, the partition method was selected, as this method requires the global system to be partitioned based on whether the displacement at certain degrees-of-freedom of the node is prescribed. All degrees-of-freedom at boundaries are marked as either essential boundary conditions or natural boundary conditions. The former prescribes the displacement of corresponding degrees-of-freedom for the nodes. If the system has *n* unknown degrees-of-freedom, the original labels of algebraic equations is reordered such that all degrees-of-freedom on the essential boundaries are grouped together. The tangential stiffness matrix and external force increment vector are assembled according to this new order, and then Equation 6.86 is recast as:

$$\begin{bmatrix} K_E & K_{EF} \\ K_{EF}^T & K_F \end{bmatrix} \begin{bmatrix} \eta_E \\ \Delta\eta_F \end{bmatrix} = \begin{bmatrix} R_E \\ \Delta F_F \end{bmatrix} \tag{6.87}$$

where η_E is the prescribed displacement vector on the essential boundaries; $\Delta\eta_F$ is the unknown displacement increment vector to be solved; R_E is the unknown reaction forces on essential boundaries and ΔF_F stands for the force increment vector exerted on the system. The second row of Equation 6.87 is rearranged as

$$K_F \Delta\eta_F = \Delta F_F - K_{EF}^T \eta_E \tag{6.88}$$

All terms on the right-hand side of Equation 6.88 are known. This linear algebraic system is efficiently solved by a variety of methods, and, the unknown displacement increment vector $\Delta\eta_F$ is computed. With the solution of $\Delta\eta_F$ and the first row of Equation 6.87, the reaction forces are found accordingly. Due to the tangential stiffness matrix generated by the frictional contact element being unsymmetrical in this study (Section 6.2), methods for symmetric coefficient matrix such as the Cholesky decomposition, conjugated gradients, etc. cannot be applied. However, many other schemes in the literature can be efficiently used to solve equation systems with an unsymmetrical coefficient matrix (Lewis et al., 2006).

6.7 Numerical Examples

6.7.1 Ultimate moment capacity of RC beams

One numerical simulation of an RC beams is presented to validate the formulation of the nonlinear RC beam elements. Frist, a simply supported RC beam with a span of 50 ft is simulated. The rectangular section is reinforced by four #8 reinforcing bars with a yielding strength of 60 ksi at an effective depth of 35 in. The concrete compressive strength is 4 ksi with a tensile strength of 472 psi. A displacement-controlled point load is applied at mid-span as shown in Figure 6.22.

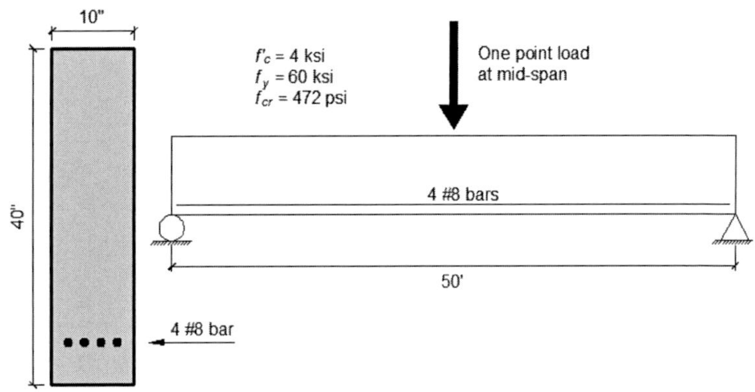

Figure 6.22 Simulated RC beam configuration

The finite element mesh consists of 20 beam elements along with 20 fibers for section discretization. This configuration yields 800 material computational points. It is enough for predicting cracking and ultimate load. The hand calculation is presented as follows:

The cracking moment and the point load at cracking are calculated as:

$$M_{cr} = \frac{f_{cr} I_g}{y_b} = \frac{472 \text{ psi} \times 5333 \text{ in.}^4}{20 \text{ in.}} = 1258667 \text{ lb-in.} \tag{6.89}$$

$$P_{cr} = \frac{4M_{cr}}{L} = \frac{4 \times 1258667 \text{ lb-in.}}{600 \text{ in.}} = 8391 \text{ lb} \tag{6.90}$$

The ultimate moment capacity and the point load at ultimate are calculated as:

$$a = \frac{A_s f_y}{0.85 f_c' b} = \frac{4 \times 0.79 \text{ in.}^2 \times 60 \text{ ksi}}{0.85 \times 4 \text{ ksi} \times 10 \text{ in.}} = 5.58 \text{ in.} \tag{6.91}$$

$$M_u = A_s f_y \left(d - \frac{a}{2}\right) = 4 \times 0.79 \text{ in.}^2 \times 60 \text{ ksi} \times \left(35 \text{ in.} - \frac{5.58 \text{ in.}}{2}\right) = 6107016 \text{ lb-in.} \tag{6.92}$$

$$P_u = \frac{4M_u}{L} = \frac{4 \times 6107016 \text{ lb-in.}}{600 \text{ in.}} = 40713 \text{ lb} \tag{6.93}$$

Table 6.2 shows the comparisons between the hand calculation and numerical simulation.

Table 6.2 Hand calculation compared to numerical simulation results

	P_{cr} (kips)	P_u (kips)
Hand Calculation	8.4	40.7
FEA	9.5	42.8
FEA/Hand Calculation	1.13	1.05

The numerical simulation agrees well with hand calculations in terms of cracking load and ultimate load capacity of the beam. Figure 6.23 plots the relation of applied load vs. mid-span deflection without reinforcement (i.e., plain concrete beam).

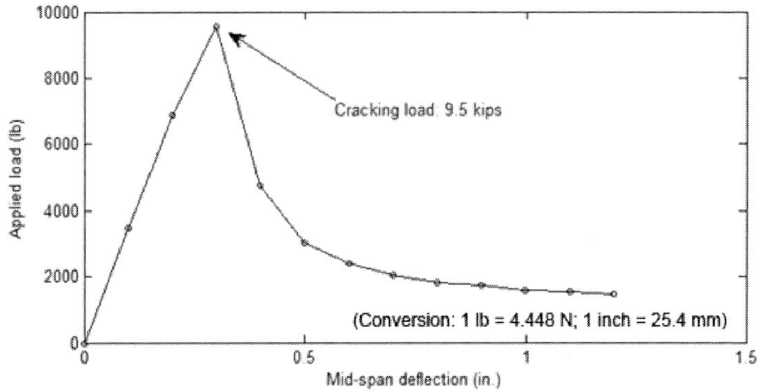

Figure 6.23 Applied load vs. mid-span deflection of a plain concrete beam

The damage pattern of the plain concrete beam is shown in Figure 6.24. Dominant cracks are developed at the mid-span region of the beam. Because cracks are assumed to be smeared along a beam element, the crack region is a little wide in the simulation. However, a more accurate local crack pattern will be obtained as the finite element mesh becomes finer.

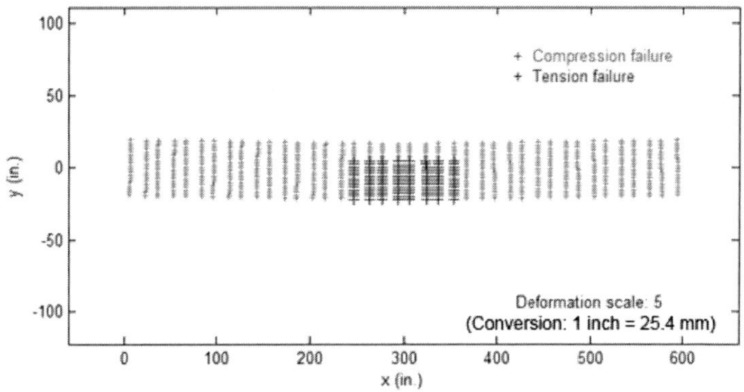

Figure 6.24 Plain concrete beam damage pattern

Figure 6.25 shows the overall relation of applied load vs. mid-span deflection with four #8 reinforcing bars added in. The applied load slightly increases as deflection increases. This might be the result of using artificial tensile residual stress according to the concrete constitutive model to maintain numerical stability. The numerical simulation has been intentionally cut off at a deflection around 7 in. Furthermore, in the current proposed method, it is difficult to predict the ultimate failure point due to the basis of finite element analysis which originates from continuum mechanics, and material fractures are hard to handle by nature.

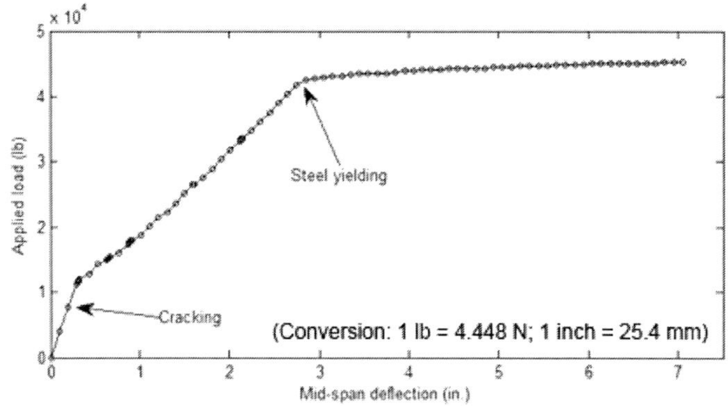

Figure 6.25 Applied load vs. mid-span deflection at ultimate state

Figures 6.26 and 6.27 show the damage pattern and strain plots of concrete fibers right at the yielding point, and Figure 6.28 plots the beam deformation (scale factor: 5) at the termination point of the simulation which clearly shows a plastic hinge developed at mid-span. The flexural type of failure mode agreed with observations from many experimental programs.

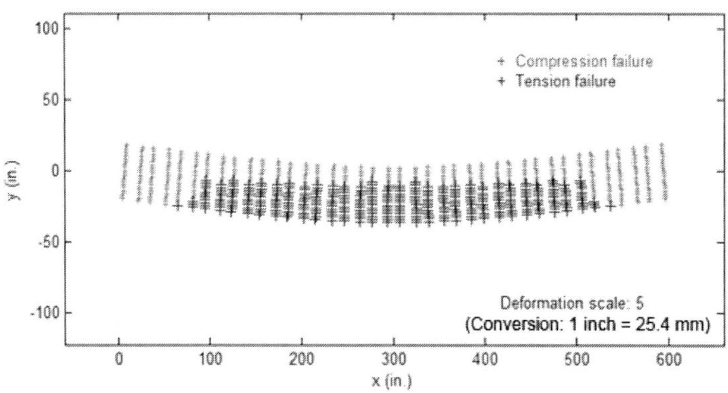

Figure 6.26 RC beam damage pattern at reinforcement yielding state

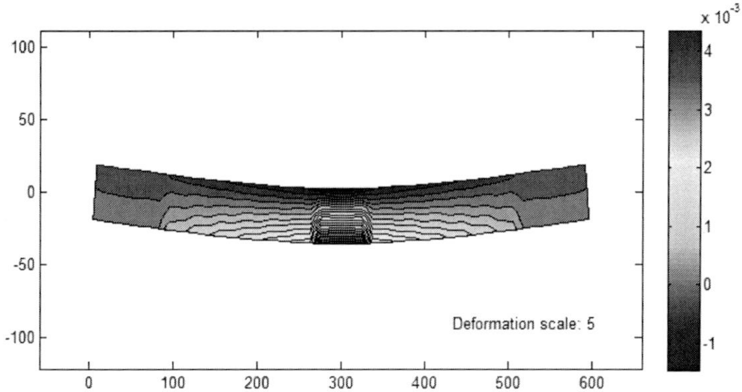

Figure 6.27 RC beam strain distribution of concrete fiber at reinforcement yielding state

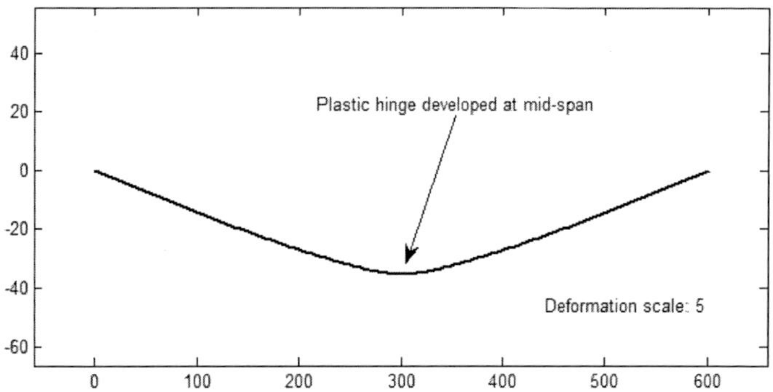

Figure 6.28 Beam deformation at termination point (deformation scale: 5)

6.7.2 Frictional loss of prestress in partially bonded PT beams

This numerical example tends to examine the performance of the formulation of frictional contact. The simulated results are compared to Code-permitted calculations. The imaginary beam is simply supported with a span of 50 ft long. The rectangular beam section has a width of 10 in. and a height of 40 in. and is reinforced only by Grade 270 seven-wire strands with an area of 1.53 in^2. The tendon profile is draped with zero eccentricity at both ends of the beam and 11 in. eccentricity at the mid-span. The tendon is unbonded and post-tensioned with an effective prestress of 204 ksi (with consideration of friction). Concrete has compressive strength of 4 ksi and tensile strength of 472 psi. The cross section and reinforcement configuration are shown in Figure 6.29

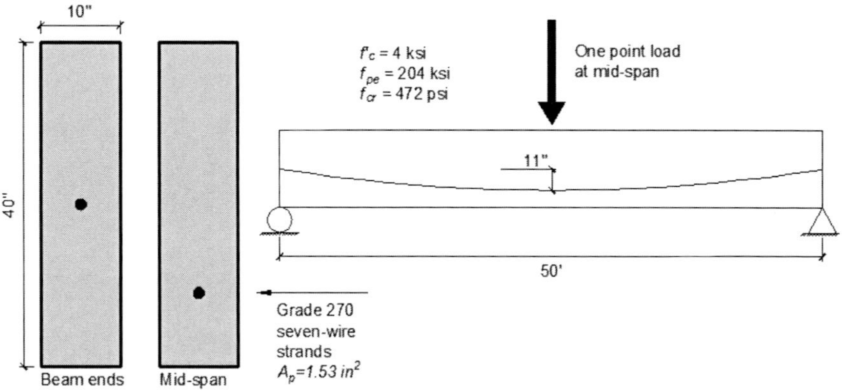

Figure 6.29 Details of simply supported beam

The finite element model contains 20 beam elements, 21 embedding elements, 21 truss elements, 20 contact elements and an anchorage element as shown in Figure 6.30.

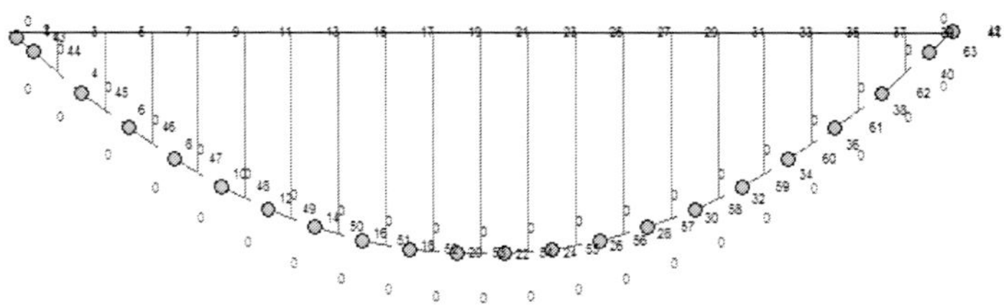

Figure 6.30 Finite element model for simply supported PT beam

A simulation is conducted of the two steps as prestressing and anchorage setting. The anchorage wedge setting was assumed to be about 0.02 in. in this case for illustrative purposes (typical wedge setting loss is much larger in practice). The frictional coefficient is assumed to be 0.3 for seven-wire strands, which yields a partially bonded interface between tendons and concrete in this simulation. The wobble effects are neglected in both simulation and hand calculation. The hand calculation is summarized first in the following paragraphs.

According to ACI 423.10R-16, the prestress at any location of the beam is computed as:

$$P_{px} = P_{pj} e^{-(Kl_{px} + \mu_p \alpha_{px})} = 204 \text{ ksi} \times e^{-0.3 \times 8 \times 11 \text{ in.}/(600 \text{ in.})^2 \times 600 \text{ in.}} = 195.2 \text{ ksi} \tag{6.94}$$

The corresponding prestress loss is computed as:

$$P_{loss} = P_{pj} - P_{px} = 204 \text{ ksi-}195.2 \text{ ksi} = 8.8 \text{ ksi} \tag{6.95}$$

The frictional loss obtained from finite element analysis (FEA) is 9.8 ksi at the right end of the beam (non-jacking end). The frictional loss is computed as a function of the longitudinal location of the beam. The ACI 423-based loss along the beam is plotted in Figure 6.31 and compared with the FEA result.

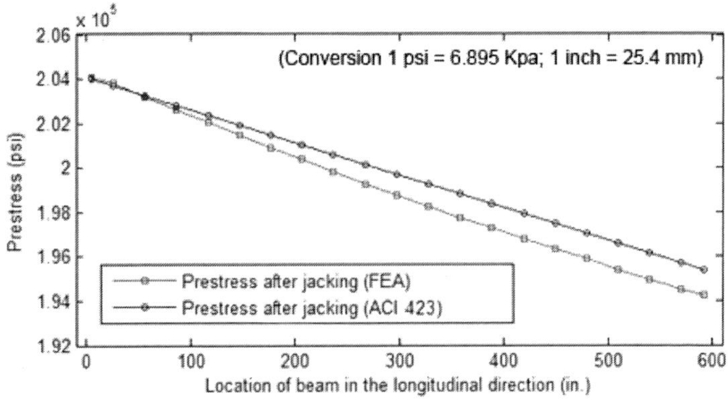

Figure 6.31 Comparison of prestress loss after jacking between FEA and ACI 423

The reverse movement of tendons due to wedge setting leads to a prestress loss at the jacking end. This wedge setting typically will result in a prestress loss at a certain distance away from the jacking end, and this distance is typically smaller than the length of the structural member. Beyond that point, there is no anchorage wedge setting loss to the prestress. This distance is a function of the anchorage wedge setting distance (slip), prestressing tendon's elastic modulus, prestressing tendon total area, and initial prestress at jacking end as well as frictional coefficient and wobble factor. The distance is explicitly computed as:

$$I_{set} = \sqrt{\frac{\Delta s A_p E_p}{\mu \alpha P_{pj}}} = \sqrt{\frac{0.02\ \text{in.} \times 1.53\ \text{in}^2 \times 28500000\ \text{psi}}{0.3 \times 8 \times 11\text{in.}/(600\ \text{in.})^2 \times 204000\ \text{psi}}} = 241.6\ \text{in.} \quad (6.96)$$

The prestress loss due to anchorage wedge setting is computed as:

$$P_{loss} = 2P_{pj}\mu\alpha I_{set} = 2 \times 204000\ \text{psi} \times 0.3 \times 8 \times 11\ \text{in.}/(600\ \text{in.})^2 \times 241.6\ \text{in.} = 3.6\ \text{ksi} \quad (6.97)$$

According to this result, the prestress loss is plotted along the beam and compared to the FEA result as shown in Figure 6.32.

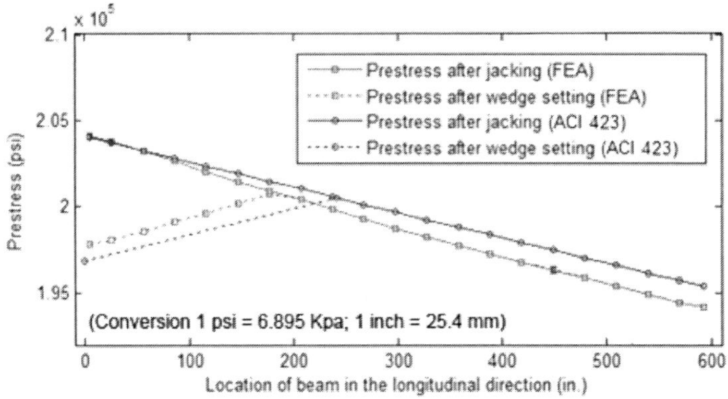

Figure 6.32 Comparison of prestress loss after anchorage wedge setting between FEA and ACI 423

Figures 6.31 and 6.32 indicate very good agreement between the finite element prediction and code based calculation, which validates the formulation of frictional contact presented in Section 6.2. The prestress losses at both ends of the beam were summarized in Table 6.3.

Table 6.3 Comparison of prestress loss

	Prestress loss at jacking end †	Prestress loss at other end ‡
Hand Calculation	3.5%	4.2%
FEA	3.0%	4.8%
FEA/Hand Calculation	0.86	1.14

† Prestress loss is computed with respect to effective prestress f_{pe} at jacking end after anchorage wedge setting
‡ Prestress loss is computed with respect to effective prestress f_{pe} at jacking end before anchorage wedge setting

6.7.3 Ultimate moment capacity of perfectly unbonded PT beam

A numerical simulation of an unbonded PT beam was conducted to investigate the performance of the truss element. The imaginary beam has the same geometry and reinforcement as the beam in the previous example in Section 6.7.2. Additionally, a point load is added at the middle span after post-tensioning to investigate flexural capacity as shown in Figure 6.33.

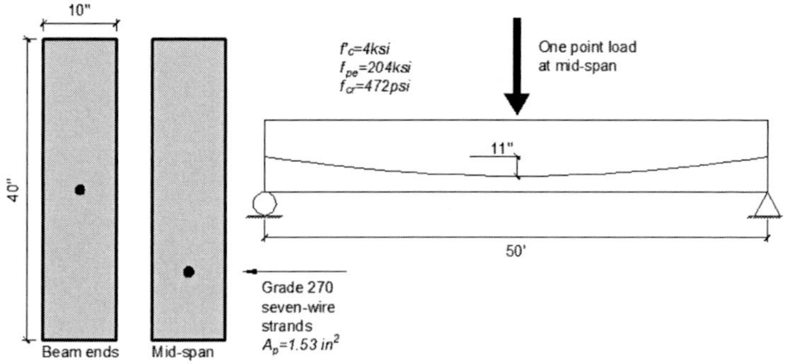

Figure 6.33 Simulated PT beam configuration

The single tendon is perfectly unbonded (frictional coefficient = 0) and post-tensioned to an effective prestress of 204 ksi. Concrete has a compressive strength of 4 ksi and a tensile strength of 472 psi. After post-tensioning, a displacement-controlled point load is applied at mid-span until the flexural failure occurs. Numerical simulation is conducted and compared with the hand calculations. The finite element model contains 20 beam elements, 21 embedding elements, 21 truss elements, 20 contact elements and an anchorage element. Simulation is conducted by three steps as prestressing, prestressing setting, and applying service load. The frictional coefficient is zero which yields a perfectly unbonded interface between tendons and concrete in this simulation. The hand calculations are summarized in the followings. Hand calculations are made in terms of prestressing camber, beam ultimate load capacity, and tendon stress at ultimate stage according to ACI 318-19 (2019) in the following paragraphs.

By the load balancing method, the equivalent balancing load intensity resulting from the pressure of the parabolic draped tendon on the concrete is calculated as:

$$w = \frac{8A_{ps}f_{ps}e}{l^2} = \frac{8 \times 1.53 \text{ in}^2 \times 204 \text{ ksi} \times 11 \text{ in.}}{(600 \text{ in.})^2} = 76.3 \text{ lb/in.} \tag{6.98}$$

The corresponding camber is calculated as:

$$\delta_{\text{camber}} = \frac{5wl^4}{384E_cI_c} = \frac{5 \times 76.3 \text{ lb/in.} \times (600 \text{ in.})^4}{384 \times 2934439 \text{ psi} \times 53333 \text{ in}^4} = 0.82 \text{ in.} \tag{6.99}$$

where E_c is the initial tangent modulus of concrete obtained from the uniaxial stress-strain relation used in the simulation.

The tendon stress at ultimate stage is calculated as:

$$\begin{aligned} f_{ps} &= f_{pe} + 10000 + \frac{f_c'}{100\rho_p} \\ &= 204\text{ksi} + 10 \text{ ksi} + \frac{4 \text{ ksi}}{100 \times 0.003825} \\ &= 224.46 \text{ ksi} \end{aligned} \tag{6.100}$$

The corresponding ultimate moment capacity and the point load at ultimate are calculated as:

$$a = \frac{A_{ps}f_{ps}}{0.85f_c'b} = \frac{1.53 \text{ in}^2 \times 224.46 \text{ ksi}}{0.85 \times 4 \text{ ksi} \times 10 \text{ in.}} = 10.1 \text{ in.} \tag{6.101}$$

$$M_u = A_{ps} f_{ps} \left(d - \frac{a}{2}\right)$$
$$= 1.53\ \text{in}^2 \times 224.46\ \text{ksi} \times \left(31\ \text{in.} - \frac{10.1\ \text{in.}}{2}\right) = 8911847\ \text{lb-in.} \quad (6.102)$$

$$P_u = \frac{4M_u}{L} = \frac{4 \times 8911847\ \text{lb-in.}}{600\ \text{in.}} = 59412\ \text{lb} \quad (6.103)$$

Table 6.4 shows the comparisons between the hand calculation and numerical simulation.

Table 6.4 Comparison between hand calculation and numerical simulation results

	δ_{camber} (in.)	P_u (kips)	f_{ps} (ksi)
Hand Calculation	0.82	59.4	224.46
FEA	0.88	56.1	226.55
FEA/Hand Calculation	1.07	0.94	1.01

The differences between the hand calculations and numerical predictions are quite small which implies the validity of the current finite element formulations for unbonded PT beams. Figure 6.34 plots the relation of applied load versus mid-span deflection. The tendon stress; f_{ps} at ultimate state of the beam is measured at the point where the applied load reaches its maximum value in the numerical test. The maximum applied load is considered as the ultimate load capacity of the simulated beam. Figure 6.35 shows the deformation of the beam (deformation scale: 5) at yielding of the mild steel bars.

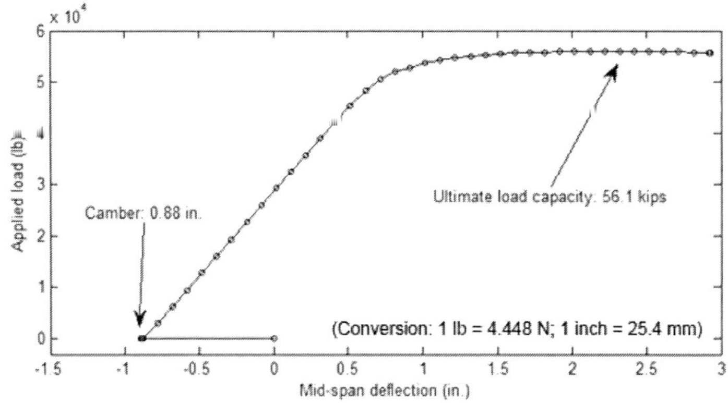

Figure 6.34 Applied load vs. mid-span deflection

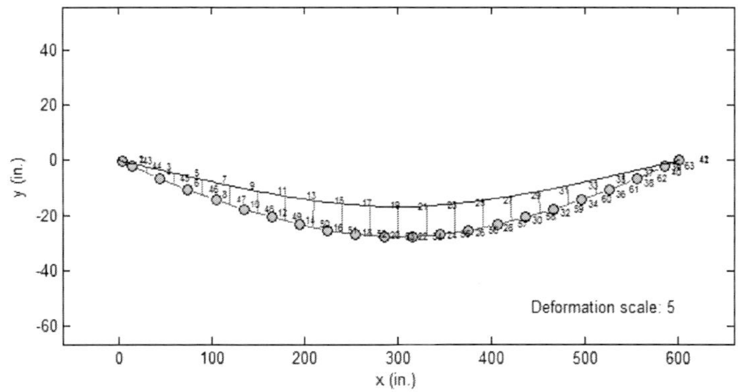

Figure 6.35 Beam deformation at reinforcement yielding (deformation scale: 5)

6.7.4 Ultimate moment capacity of unbonded PT beams with consideration of friction

This example allows comparison of the difference between the imaginary PT beam with and without consideration of friction between tendons and corresponding sheathings. All parameters are the same as in the example in Section 6.7.3, except that a friction coefficient of 0.3 is assumed and added to the model. The beam is post-tensioned to an effective prestress of 204 ksi measured at the jacking end. Anchorage setting loss is not considered in accordance with the previous example.

The single point displacement load at mid-span is used to investigate the ultimate moment capacity. The relation of load displacement is plotted in Figure 6.36. The strain contours of concrete fibers are plotted immediately after post-tensioning and at ultimate stage (Figures 6.37, 6.38 and 6.39).

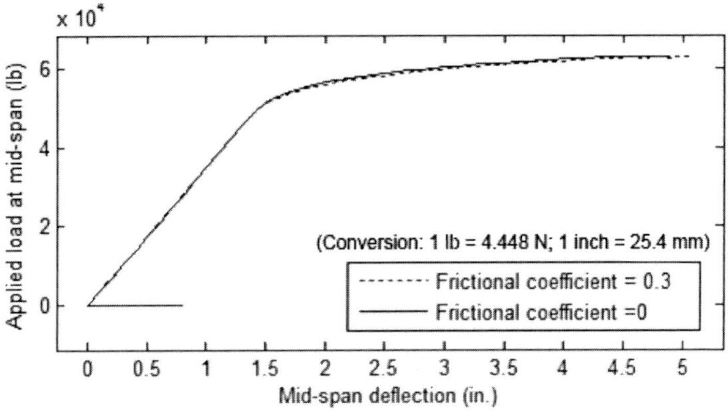

Figure 6.36 Comparison of global response between partial bonding and perfect bonding

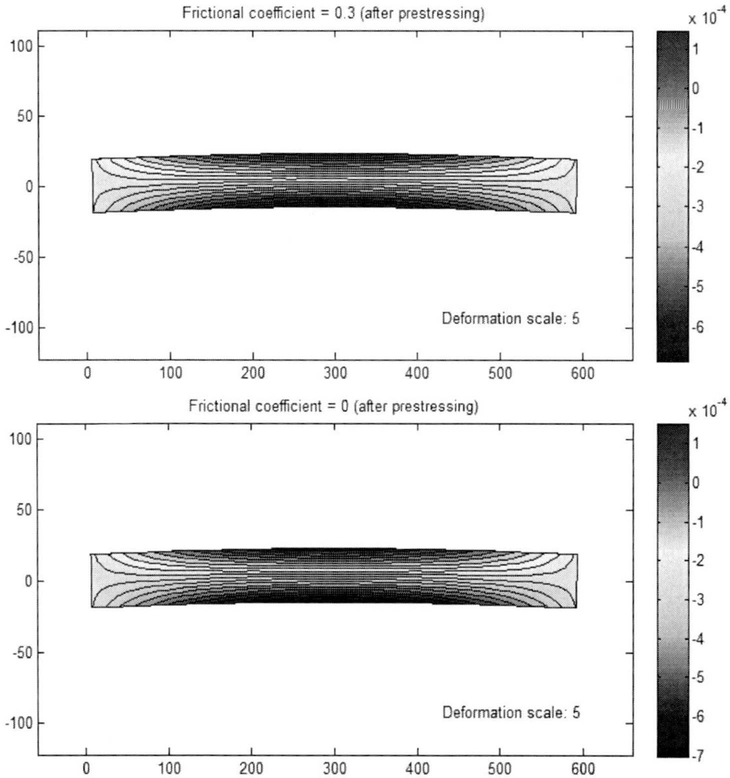

Figure 6.37 Strain contours at post-tensioning stage

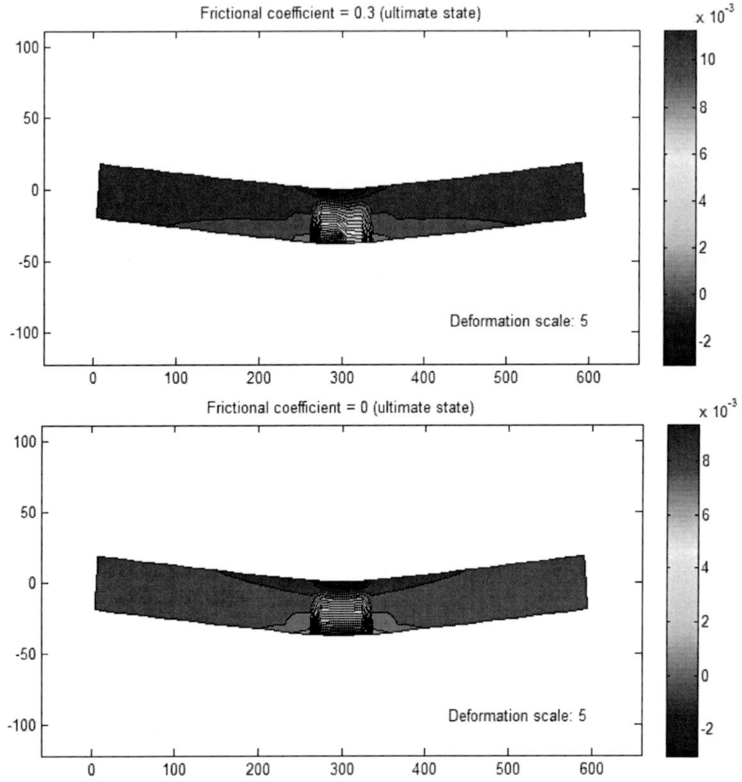

Figure 6.38 Strain contours at ultimate stage

The difference is negligible in the above figures, indicating that the frictional effect is neglected in terms of global response. The crack patterns obtained from both simulations show a similar style (Figure 6.39).

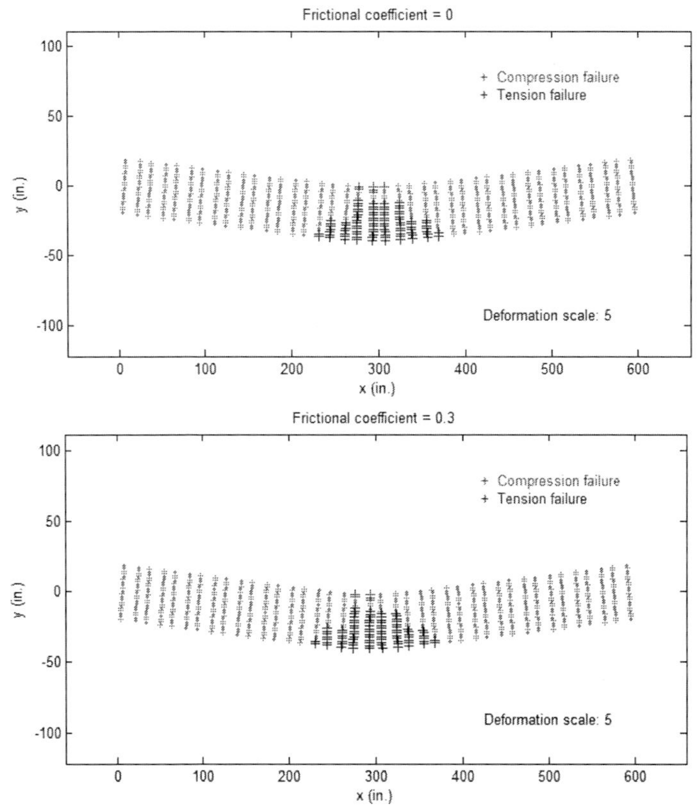

Figure 6.39 Comparison of failure patterns with and without friction

However, the local responses of prestressing steel are different at ultimate. Figure 6.40 shows the comparison of strain and stress of prestressing steel at ultimate where the prestress tends to be larger for the case of perfect bonding. Even at the mid-span section, the prestress with consideration of friction is slightly smaller than its counterpart with perfect bonding condition. The greatest difference occurs at the both ends of the beam. This explains a phenomenon observed in the previous finite element analyses in Chapter 6 where the numerically predicted prestress is always larger than the observed values in the experiments. The reason behind the difference is the frictional effect.

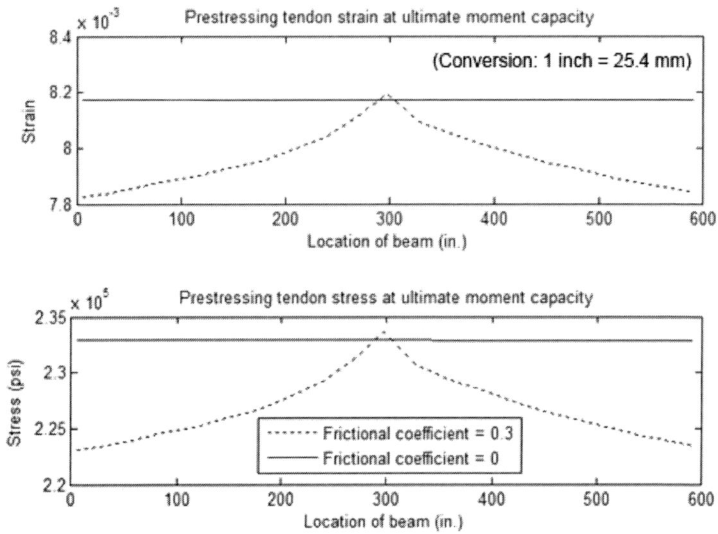

Figure 6.40 Comparison of prestressing tendon strain and stress at ultimate moment capacity

6.7.5 Numerical simulation of one-way unbonded PT slabs

This example aims to conduct numerical simulations of Slab4 and Slab5 which were tested by Cooke et al. (1981). The global responses were compared to both numerical simulations and experiments presented in Chapter 5. The description of specimens and test configuration were described in Section 5.5. Since global responses are barely influenced by the tendon bonding, and Slab4 and Slab5 have a relatively short span along with a straight tendon profile, the prestress loss was neglected in this example.

Figure 6.41 and 6.42 show the comparisons of global responses of Slab4 and Slab5, respectively. The current finite element formulation predicted a perfect load-displacement relation for Slab4 while overestimated moment capacity of Slab5 by 11% compared to experiments. The possible explanation is that the reinforcement ratio of Slab5 is only about half that of Slab4, which resulted in a more concentrated cracking pattern against a distributed cracking pattern observed from Slab4 in the experiments. The current beam element formulation presented in Section 6.5 assumed smeared cracking along the beam element. Large and concentrated plastic hinge may not be accurately captured in the current simulation.

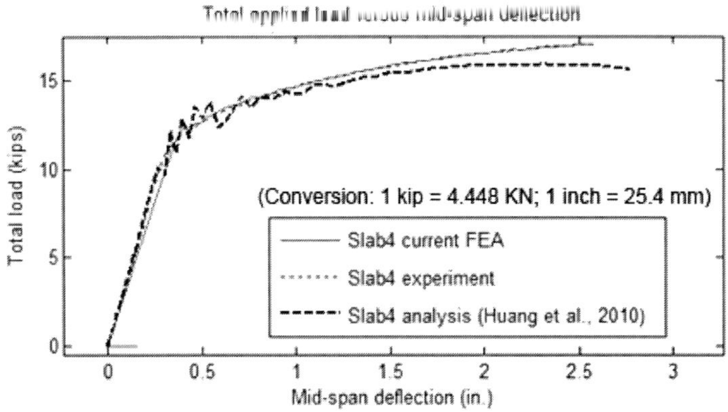

Figure 6.41 Comparison of global response of Slab4

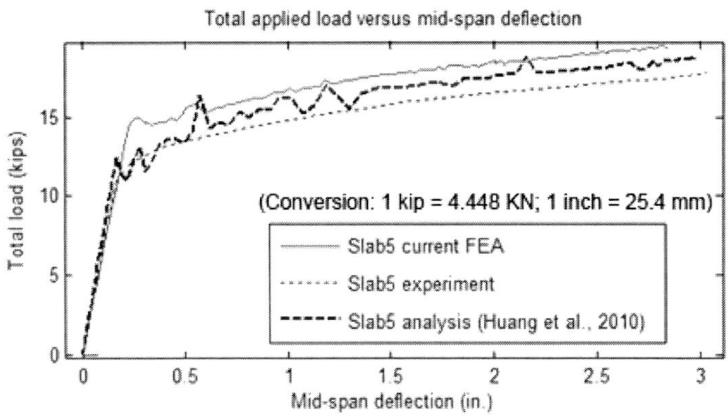

Figure 6.42 Comparison of global response of Slab5

Table 6.5 provides the summary of applied load at the experimental termination point of mid-span deflection.

Table 6.5 Comparison of applied load (kips) at experimental termination point of mid-span deflection

	Slab4	Slab5
Current FEA (kips)	17.07	19.44
Chapter 5 analysis (kips)	15.85	18.71
Experiment (kips)	17.05	17.51
Current FEA/Experiment (%)	1.001	1.11

6.8 Summary

This chapter presented a general frame work of nonlinear finite element formulation towards simulating two-dimensional PT frames. The nonlinear natures of inelastic behavior due to material nonlinearity and large displacement were addressed in the formulation. Furthermore, the contact interface between tendons and corresponding sheathings are formulated directly via NTS discretization. Various elements were developed to assemble a complete PT system. The beam elements were used to simulate individual frame members. The Euler-Bernoulli type beam element was implemented in this study for the application of slender beam problems and for demonstration purpose. It required trivial work to implement other beam elements with higher order theory. The post-tensioning tendon was assembled via nonlinear truss element. Corresponding tendon sheathing was embedded into the parent frame member through the embedding element. The contact element forms the interface between the tendons and corresponding tendon sheathings. Contact constraints were enforced by the penalty form with the ability of considering both frictionless and frictional contacts. A special element dealing with PT tendon anchorage system was developed imitating the process of prestressing in practice. The Newton-Raphson with line search modification was adopted as the numerical algorithm solving time independent problems. All formulations were implemented and programmed in MATLAB (2010b). The reliability and robustness of the proposed formulations were validated through several numerical examples carried out in MATLAB.

CHAPTER SEVEN

CONCLUSION

In this book, the finite element modeling schemes and corresponding applications to bonded and unbonded PT structures have been discussed in depth. Part one of this book discussed approaches of modeling PT structures via general purpose finite element packages. Among many excellent finite element packages, ABAQUS was selected as the platform to develop the numerical models for the following researches. Chapter 4 presents the backgrounds of modeling techniques and corresponding theories used to assemble a complete finite element model. The preferences for material constitutive models (concrete and steel), element library, concrete-steel interaction technique, and solving algorithm are thoroughly discussed in Chapter 4. In particular, the research was zeroed in on the great challenge of modeling interactions between concrete and prestressing tendons. Two approaches were proposed, including the traditional spring system method and the more advanced contact formulation. Followed by a series of numerical studies in Chapter 5, the modeling methodology was well validated against experimental data. Three practical engineering problems were investigated through the numerical studies including 1) punching shear failure of two-way PT slabs; 2) stress increment in prestressing tendons at the failure of structural members; and 3) the influence of different PT systems (i.e., bonded PT vs. unbonded PT) on behaviors of PT structures. Some drawbacks of the generic modeling approaches were observed. Particularly, the desire of more efficient and accurate solution process and modeling flexibility motivated the materials of part two in this book (Chapter 6).

In Chapter 6, an innovative nonlinear finite element formulation for two-dimensional PT frames was developed to overcome the difficulties presented by the generic modeling schemes via general purpose finite element packages. The proposed formulation incorporated five different elements (most are engineering type) to assemble a complete PT frame system. The most critical developments lay on the contact element with finite sliding capability. In addition, an anchorage element was proposed to simplify the complicated prestressing procedure in practice. Other traditional nonlinear beam and truss elements were also presented in detail in Chapter 6. A series of numerical examples conducted at the end of this chapter validated the robustness of the formulation. The solutions were improved compared with the modeling approaches proposed in Chapter 5 in terms of efficiency and accuracy.

Based on the discussions, analyses, and developments through Chapter 4 to Chapter 6, the following verdicts were drawn with respect to the proposed finite element modeling schemes.

(1) In the context of modeling via general purpose finite packages, the interaction between prestressing tendons and corresponding sheathings can be well simulated by either approximating the sliding behavior (spring method) or directly formulating contact interface in the numerical model (contact approach). Typically, the spring method yields cost-efficient modeling where the efficiency of solution is greatly valued. However, it is limited to structures with thick concrete coverage over prestressing tendons. On the other hand, the computationally anxious contact approach has no limitations of the minimum concrete coverage and provides better modeling flexibility. The preference of the interaction models highly depends on individual problems.

(2) The proposed modeling schemes based on general purpose finite element packages tend to slightly overestimate stress in prestressing tendons at the limit state of structures. The possible cause might be the simplification assumed for the process of post-tensioning where frictional effects are neglected. The prestress loss, however, is observed in practice, especially for structures prestressed with long continuous prestressing tendons. It is suggested to apply the proposed model to large-scale PT structures with caution if frictional effects are left off.

(3) The proposed finite element modeling schemes via general purpose finite element packages are reliable and robust. It performed well to predict global flexural behaviors of various types of PT structures. However, there is still room for improvement regarding predictions of local behaviors and overall shear behaviors. Apparently, the computational anxiety becomes a big hurdle preventing the model from applying to large-scale structures. The modeling scheme based on general purpose finite element packages is mostly suitable and efficient for evaluating and analyzing a single PT structural member in substantial details.

(4) The proposed two-dimensional nonlinear finite element formulations for PT frames and corresponding implementations in MATLAB were validated to be very efficient and robust through a series of numerical simulations. The incorporation of the contact technique with engineering-type elements was successful. Particularly, improvement

towards the prediction of the local behavior of prestressing tendons was accomplished in comparison with the modeling scheme on the basis of general purpose finite element packages. The largest merit of the proposed formulation is its computational friendliness, which is of great value for modelling large-scale PT structures. The proposed formulation is appropriate to aid the design and limit state analysis for PT frames.

In addition to the discussions of finite element formulations regarding PT structures, investigations of three practical engineering problems were conducted based on the proposed FEM. The corresponding evaluations of the current building code, ACI 318, innovatively incorporated the numerical analyses and experimental data. The evaluation processes showed the efficiency of analyzing numerical data on an experimental basis. Many inaccessible data in experiments can be accessed in finite element models. Therefore, the assessments of ACI 318-19 were conducted accordingly in Chapter 5. The conclusions of the evaluations are summarized as follows:

(1) In the case of exterior and corner slab-column connections in two-way PT slabs, the unbalanced moment transfer ratio γ_v permitted in ACI 318-19 is suggested to be relieved. In addition, the numerical simulations clearly indicated that the punching shear capacity V_c benefits from prestressing, even though the specimens were not permitted to include prestressing in the calculations of punching shear capacity as per ACI 318-19. More investigations are needed to evaluate and quantify the prestress effect on the punching shear capacity for PT slab-column connections.

(2) In the case of corner slab-column connections in two-way PT slabs, the assessment of the eccentric shear stress model (ACI 318-19) indicated that the model is reasonably reliable for predicting shear redistribution on the periphery of slab-column connections with negligible to moderate initial prestress. It is recommended to include the initial prestress effect in the eccentric shear stress model if the slab has a large initial precompression force around the slab-column connections.

(3) In the case of interior slab-column connections in two-way PT slabs, negligible difference was found between the bonded tendon systems and unbonded tendon systems in terms of flexural behavior. The shear redistributions on the periphery of the interior slab-column connection also showed similar behaviors in both PT systems. In the case of one-way PT slabs and beams designed under the guidelines of the current building code, the bonding condition of prestressing tendons barely influences the flexural behavior of structural members. The ultimate moment capacities were similar in both cases according to numerical simulations and

experiments. Therefore, it is concluded that some types of PT structures behave similarly in terms of flexural and shear performance.

REFERENCES

ABAQUS, "ABAQUS User's Manual, Volumes I to III," Pawtucket, Rhode Island, 2003.

Aroni, S., "Strength of Slender Prestressed Concrete Columns," *Prestressed Concrete Institute – Journal*, Vol. 13, No. 2, 1968, pp. 19-33.

ACI Committee 318, "Building Code Requirements for Reinforced Concrete (ACI 318-63)" ACI, Detroit, Michigan, 1963, 144 pp.

ACI Committee 318, "Building Code Requirements for Reinforced Concrete (ACI 318-77)" ACI, Detroit, Michigan, 1977, 103 pp

ACI Committee 318, "Building Code Requirements for Structural Concrete (ACI 318-19)," ACI, Farmington Hills, Michigan, 2019, 988 pp.

Breckenridge, R. A., and Bugg, S. L., "Effects of Long-Time Loads on Prestressed Concrete Beams," *Prestressed Concrete Institute – Journal*, Vol. 9, No. 6, 1964, pp. 75-89.

Brotchie, J. F. and Beresford, F. D., "Experimental Study of Prestressed Concrete Flat Plate Structure," *Civil Enginerring Transactions*, Institution of Engineers, Australia, Vol. 9, No. 2, 1967, pp. 276-282.

Burns, N. H. and Hemakom, R., "Test of Scale Model Post-Tensioned Flat Plate," *Journal of the Structural Division, ASCE*, Vol. 103, No. 6, June 1977, pp. 1237-1255.

Burns, N. H. and Hemakom, R., "Test of Post-Tensioned Flat Plate With Banded Tendons," *Journal of Structural Engineering, ASCE*, Vol. 111, No. 9, 1985, pp. 1899-1915.

Cooke, N.; Park, R., and Yong, P., "Flexural Strength of Prestressed Concrete Members with Unbonded Tendons," *Prestressed Concrete Institute*, Vol. 26, No. 6, 1981, pp. 52-80.

Carreira, D. J., and Chu, K. H., "Stress-Strain Relationship for Plain Concrete in Compression," *ACI Journal, Proceedings*, Vol. 83, No. 6, 1985, pp. 797-804.

Darwin, D., and Pecknold, D. A., "Analysis of RC Shear Panels under Cyclic Loading," *Journal of Structural Engineering, ASCE*, Vol. 102, No. 2, 1976, pp. 355-369.

Darwin, D., and Pecknold, D. A., "Nonlinear Biaxial Stress-Strain Law for Concrete," *Journal of Structural Engineering, ASCE*, Vol. 103, No. 2, 1976, pp. 229-241.

Devalapura, R. K., and Tadros, M. K., "Critical Assessment of ACI 318 Eq. (18-3) for Prestressing Steel Stress at Ultimate Flexure," *ACI Structural Journal*, Vol. 89, No. 5, 1992, pp. 538-546.

Drucker, D. C., and Prager, W., "Soil Mechanics and Plastic Analysis or Limit Design," *Quarterly of Applied Mathematics*, Vol. 10, No. 2, 1952, pp. 157-165.

El-Mezaini, N.; Balkaya, C.; and Çitipitioğlu, E., "Analysis of Frames with Nonprismatic Members," *Journal of Structural Engineering, ASCE*, Vol. 117, No. 6, 1991, pp. 1573-1592.

Ellobody, E. A., and Bailey, C. G., "Modelling of bonded post-tensioned concrete slabs in fire." *Proceedings of ICE, Structures and Buildings*, Vol. 161, No. 6, 2008, pp. 311-323.

Foutch, D. A.; Gamble, W. L., and Sunidja, H., "Tests of Post-Tensioned Concrete Slab-Edge Column Connections," *ACI Structural Journal*, Vol. 87, No. 2, 1990, pp. 167-179.

Fish, J., and Belytschko, T., "A First Course in Finite Elements," Wiley, 2007, 336 pp.

Gamble, W.L., "An Experimental Investigation of The Strength And Behaviour of A Prestressed Concrete Flat Plate," Report T80-9, Division of Building Research, Commonwealth Scientific and Industrial Research Organization, Melbourne, Australia, 1964.

Gardner, N. J., and Kallage, M. R., "Punching Shear Strength of Continuous Post-Tensioned Concrete Flat Plates," *ACI Materials Journal*, Vol. 95, No. 3, 1998, pp. 272-283.

Galerkin, B. G., "Rods and Plates: Series Occurring in Various Questions Concerning the Elastic Equilibrium of Rods and Plates," *Engineers Bulletin (Vestnik Inzhenerov)*, Vol. 19, 1915, pp. 897–908 (in Russian).

Huang, Y.; Kang, T. H.-K.; Ramseyer, C.; and Rha, C., "Background to Multi-Scale Modeling of Unbonded Post-Tensioned Concrete Structures," *International Journal of Theoretical and Applied Multiscale Mechanics*, Vol. 1, No. 3, 2010, pp. 219-235.

Joint ACI-ASCE Committee 423, "Guide to Estimating Prestress Loss," ACI, Farmington Hills, Michigan, 2016, 64 pp.

Kang, Y.-J., and Scordelis, A. C., "Nonlinear Analysis of Prestressed Concrete Frames," *Journal of Structural Engineering*, *ASCE*, Vol. 106, No. 2, 1980, pp. 445-462.

Kosut, G. M.; Burns, N. H.; and Winter, C. V., "Test of Four-Panel Post-Tensioned Flat Plate," *Journal of Structural Engineering*, *ASCE*, Vol. 111, No. 9, 1985, pp. 1916-1929.

Kang, T. H.-K., and Wallace, J. W., "Stresses in Unbonded Tendons of Post-Tensioned Flat Plate Systems under Dynamic Excitation," *PTI Journal*, Vol. 6, No. 1, 2008, pp. 31-44.

Lin, T. Y., "Strength of Continuous Prestressed Concrete Beams under Static and Repeated Loads," *ACI Journal*, *Proceedings*, Vol. 26, No. 10, 1955, pp. 1037-1059.

Lee, J., and Fenves, G. L., "Plastic-Damage Model for Cyclic Loading of Concrete Structures," *ASCE Journal of Engineering Mechanics*, Vol. 124, No. 8, 1998, pp. 892-900.

Lubliner, J.; Oliver, J.; Oller, S.; and Onate, E., "Plastic-Damage Model for Concrete," *International Journal Solids and Structures*, Vol. 25, No. 3, 1989, pp. 299-326.

Lewis, J. M.; Lakshmivarahan, S.; and Dhall, S. K., "Dynamic Data Assimilation: A Least Squares Approach," Cambridge University Press, 2006, 680 pp.

MATLAB, version 7.11 (R2010b), Natick, Massachusetts: The MathWorks Inc., 2010.

Mattock, A. H.; Yamazaki, J.; and Kattula, B. T., "Comparative Study of Prestressed Concrete Beams, With and Without Bond," *ACI Journal, Proceedings*, Vol. 68, No. 2, 1971, pp. 116-125.

Martinez-Cruzado, J. A., "Experimental Study of Post-Tensioned Flat Plate Exterior Slab-Column Connections Subjected to Gravity and Biaxial Loading," Ph.D. dissertation, University of California, Berkeley, 1993.

Nikolic, Z., and Mihanovic, A., "Non-Linear Finite Element Analysis of Post-Tensioned Concrete Structures," *Engineering Computations*, Vol. 14, No. 5, 1997, pp. 509-528.

Ngo, D., and Scordelis, A. C., "Finite Element Analysis of Reinforced Concrete Beams," *ACI Journal, Proceedings*, Vol. 64, No. 3, 1967, pp. 152-163.

Odello, R. J., and Mehta, B. M., "Behavior of A Continuous Prestressed Concrete Slab with Drop Panels," M.S. thesis, Division of Structural Engineering and Structural Mechanics, University of California, Berkeley, 1967.

Pimanmas, A.; Warnitchai, P.; and Pongpornsup, S., "Seismic Performance of 3/5 Scaled Post-Tensioned Interior Flat Slab- Column Connections," Proceedings of Asia Conference on Earthquake Engineering (ACEE2004), 5-6 March 2004, Manila, Phillipines.

Prawatwong, U.; Tandian, C.H.; and Warnitchai, P., "Seismic Performance Enhancement of Post-Tensioned Flat Plate System with Drop Panel," 8th Pacific Conference on Earthquake Engineering (8PCEE), Singapore, 2007.

Park, H.-G., and Choi, K.-K., "Strength of Exterior Slab-Column Connections Subjected to Unbalanced Moments," *Engineering Structures, Elsevier*, Vol. 29, June 2007, pp. 1096-1114.

Park, R., and Gamble, W. L., "Reinforced Concrete Slabs," John Wiley and Sons, Inc., Second Edition, New York, NY, 2000, 736 pp.

Pőttler, R., and Swoboda, G. A., "Nonlinear Beam Element for RC Structures," *Communications in Applied Numerical Method*, Vol. 3, No. 5, 1987, pp. 397-406.

Scordelis, A. C.; Lin, T. Y.; and Itaya, R., "Behavior of Continuous Slab Prestressed in Two Directions," *ACI Journal*, *Proceedings*, Vol. 56, No. 12, December 1959, pp. 441-460.

Squire, W., and Trapp, G. (1998), "Using Complex Variables to Estimate Derivatives of Real Functions," *SIAM Review,* Vol. 40, No. 1, pp. 119-112

Stavroulaki, M. E.; Leftheris, B. P.; and Stavroulakis, G. E., "Optimal Prestress in Modal Analysis via Induced Temperature Modeling," *Structural Optimization*, Vol. 13, No. 2-3, 1997, pp. 95-103.

Stravroulaki, M. E.; Stavroulakis, G. E.; and Leftheris, B., "Modeling Prestress Restoration of Buildings by General Purpose Structural Analysis and Optimization Software, the Optimization Module of MSC/NASTRAN," *Computers and Structures*, Vol. 62, No. 1, 1997, pp. 81-92.

Smith, S. W., and Burns, N. H., "Post-Tensioned Flat Plate to Column Connection Behavior," *PCI Journal*, Vol. 19, No.3, May-June 1974, pp. 74-91.

Trongtham, N., and Hawkins, N. M., "Moment Transfer to Columns in Unbonded Post-Tensioned Prestressed Concrete Slabs," Report SM77-3, Department of Civil Engineering, University of Washington, Seattle, WA, 1977, 186 pp.

Van Greunen, J., and Scordelis, A. C., "Nonlinear Analysis of Prestressed Concrete Slabs," *Journal of Structural Engineering, ASCE*, Vol. 109, No. 7, 1983, pp. 1742-1760.

Vecchio, F. J.; Gauvreau, P.; and Liu, K., "Modeling of Unbonded Post-Tensioned Concrete Beams Critical in Shear," *ACI Structural Journal*, Vol. 103, No. 1, 2006, pp. 57-64.

Warnitchai, P.; Pongpornsup, S.; and Prawatwong, U., "Seismic Performance of Post-Tensioned Interior Flat Slab-Column Connections," Proceedings of Third International Symposium on New Technologies for Urban Safety of Mega Cities in Asia, India, 2004.

Wriggers, P., "Nonlinear Finite Element Methods," 1st edition, Springer, 2008, 572 pp.

Yang, K.-H., and Kang, T. H.-K., "Equivalent Strain Distribution Factor for Unbonded Tendon Stress at Ultimate," *ACI Structural Journal*, Vol. 108, No. 2, 2011, pp. 217-226.

Zienkiewicz, O. C., and Taylor, R. L., "The Finite Element Method for Solid and Structural Mechanics, Sixth Edition (Volume 2)," Butterworth-Heinemann, 2005, 736 pp.

APPENDIX

MATLAB SOURCE CODES

The main program contains 8 class objects and 40 sub-routines. A pre-processing module and 3 post-processing modules were also developed to help quickly process examples in the study. All source codes were debugged in MATLAB (2010b) and shown below.

Classes define:

ClassAnchorElement.m

```
% Define anchorage element class
classdef ClassAnchorElement
  properties
    Nodes;                    % Nonde labels within the element
    Epsilon;                  % Penalty stiffness factor
  end
end
```

ClassBeamElement.m

```
% Define Beam element class
classdef ClassBeamElement
  properties
    Nodes;                    % Nonde labels within the element
    Length;                   % Element length
    A;                        % Cross section area
    S;                        % Firs moment of ineratia
    I;                        % Secondary moment of ineratia
    hf;                       % Distance between ref axis and each fiber
centorid
    n;                        % Number of fiber sections
    As;                       % Area of steel of each layer e.As =
[layer1,layer2..]
    Ss;                       % Firs moment of ineratia of steel layer
    Is;                       % Second moment of ineratia of steel layer
```

```
    zs;                      % Distance between each steel layer and ref
axis
  end
end
```

ClassContactElement.m

```
% Define Contact element class
classdef ClassContactElement
  properties
    Nodes;                   % Nonde labels within the element
    Epsilon;                 % Penalty stiffness factor
  end
end
```

ClassEmbeddingElement.m

```
% Define Embedding element class
classdef ClassEmbeddingElement
  properties
    Nodes;                   % Nonde labels within the element
    Length;                  % Element length
    A;                       % Cross section area
    E;                       % Youngs' modulus
    I;                       % Secondary moment of ineratia
  end
end
```

ClassMaterial.m

```
% Define material library class
classdef ClassMaterial
  properties
    Mu;                      % Coulomb's frictional coefficient
    Concrete;                % Concrete unaxial stress-strain data
    Steel;                   % Steel unaxial stress-strain data
    Psteel;                  % Prestressing steel unaxial stress-strain data
    Ec;                      % Concrete Young's modulus array
    Ect;                     % Concrete tangential stiffness array
    Es;                      % Steel Young's modulus array
    Est;                     % Steel tangential stiffness array
```

```
    Eps;                    % Prestressing steel stiffness array
    Epst;                   % Prestressing steel tangential stiffness array
  end
end
```

ClassNode.m

```
% Define node class
classdef ClassNode
  properties
    X;                      % Position at global X axis
    Y;                      % Position at global Y axis
    U;                      % Displacement at global X axis
    V;                      % Displacement at global Y axis
    T;                      % Slop at global coordinate
    Fx_residual;            % Residual force at global X axis
    Fy_residual;            % Residual force at global Y axis
    M_residual ;            % Residual Moment at global coordinate
    Boundary;               % Flag of essential and natural boundary
conditions
    DOF;                    % Degrees of freedom
  end
end
```

ClassODB.m

```
% Define Model Database
classdef ClassODB
  properties
    Node;
    BeamElement;
    TrussElement;
    AnchorElement;
    ContactElement;
    EmbeddingElement;
    nd;                     % Number of degrees-of-freedom on the
essential boundary
    nnd;                    % Number of nodes in golbal system
    neq;                    % Number of equations
    ID;                     % Nodes mapping1
    ID2;                    % Nodes mapping2
```

```
    d_E;                    % Reordered essential boundary
    anchor;                 % 0--no anhor  1--anhor stage
    anchor_u;               % relative u_disp between anchor nodes before
prestressing
    anchor_v;               % relative v_disp between anchor nodes before
prestressing
    slip;                   % node-slip array (plastic displacement due to
slip)
    Material;               % Material class for material parameters
  end
end
```

ClassTrussElement.m

```
% Define Truss element class
classdef ClassTrussElement
  properties
    Nodes;                  % Nonde labels within the element
    Length;                 % Element length
    A;                      % Cross section area
    E;                      % Youngs' modulus
  end
end
```

Functions:

Assembly_F_aci.m

```
% compute anchorage residual force vector
function F_ac = Assembly_F_aci(epsilon,x,y,u,v,ID,ID2,eleac,neq)

F_ac = zeros(neq,1);        % Initialize contact residual from normal
contact
n1 = [1;0;-1;0;0];
n2 = [0;1;0;-1;0];
for i = 1 : size((eleac),1)

  % current distance between two nodes substracting displace before
  % prestressing stage
  hx = x(eleac(i,2))-x(eleac(i,1))-u(i);
  hy = y(eleac(i,2))-y(eleac(i,1))-v(i);
  F_cie = epsilon*(hx*n1+hy*n2);
```

```
  a = ID(ID2(eleac(i,1),1));
  b = ID(ID2(eleac(i,1),2));
  c = ID(ID2(eleac(i,2),1));
  d = ID(ID2(eleac(i,2),2));
  e = ID(ID2(eleac(i,2),3));
  temp = [a,b,c,d,e];
  for n = 1:5
    F_ac(temp(n)) = F_cie(n);
  end
end
```

```
Assembly_F_ci.m
% compute contact residual force vector from normal and tangential
contact
function [F_ci,slip2] =
Assembly_F_ci(epsilon,x,y,xi,yi,ID,ID2,elec,neq,slip,U,Mu)

slip2 = slip;      % slip2 store the updated tendon slip at end of each
iteration
F_ci = zeros(neq,1);   % Initialize contact residual from normal contact
for i = 1 : size((elec),1)

  % current lengeth of one element
  l = sqrt((y(elec(i,3))-y(elec(i,2)))^2+(x(elec(i,3))-x(elec(i,2)))^2);

  % unit tangent vector
  t(1) = (x(elec(i,3))-x(elec(i,2)))/l;
  t(2) = (y(elec(i,3))-y(elec(i,2)))/l;

  % unit normal vector
  n(1) = -t(2);
  n(2) = t(1);

  Xi = 1/l*[x(elec(i,1))-x(elec(i,2)),y(elec(i,1))-y(elec(i,2))]*t';
  Xi0 = 1/l*[xi(elec(i,1))-xi(elec(i,2)),yi(elec(i,1))-yi(elec(i,2))]*t';
  Ns = [n,-(1-Xi)*n,0,-Xi*n,0,0,0,0,0]';
  N0s = [0,0,-n,0,n,0,0,0,0,0]';
  Ts = [t,-(1-Xi)*t,0,-Xi*t,0,0,0,0,0]';
  T0s = [0,0,-t,0,t,0,0,0,0,0]';
```

```
% unit middle vector between two truss element where contact master
% node is in the middle
n1(1) = (xi(elec(i,4))-xi(elec(i,1)))/sqrt((xi(elec(i,4))-
xi(elec(i,1)))^2+(yi(elec(i,4))-yi(elec(i,1)))^2)/2+(xi(elec(i,5))-
xi(elec(i,1)))/sqrt((xi(elec(i,5))-xi(elec(i,1)))^2+(yi(elec(i,5))-
yi(elec(i,1)))^2)/2;
n1(2) = (yi(elec(i,4))-yi(elec(i,1)))/sqrt((xi(elec(i,4))-
xi(elec(i,1)))^2+(yi(elec(i,4))-yi(elec(i,1)))^2)/2+(yi(elec(i,5))-
yi(elec(i,1)))/sqrt((xi(elec(i,5))-xi(elec(i,1)))^2+(yi(elec(i,5))-
yi(elec(i,1)))^2)/2;

%if tendon is complete straight, n1 vector will become [0,0]. In
%that case, n1 will be corrected as a vector orthgonal to either of
%the truss element vectors
if n1 == [0,0]
   n1(1)=-(yi(elec(i,4))-yi(elec(i,1)));
   n1(2)=xi(elec(i,4))-xi(elec(i,1));
end

n1=n1/norm(n1,2);
tan_beta = 1*abs(n1(1)*n(2)-n1(2)*n(1))/(n1(1)*n(1)+n1(2)*n(2));

%compute correction force unit directional vector t1
if (abs( n1(1)*t(1)+n1(2)*t(2))) == 0
   t1 = 0;
else
   t1 = -(n1(1)*t(1)+n1(2)*t(2))/(abs( n1(1)*t(1)+n1(2)*t(2)));
end

gns = dot([x(elec(i,1));y(elec(i,1))]-(1-Xi)*[x(elec(i,2));y(elec(i,2))]-
Xi*[x(elec(i,3));y(elec(i,3))],n);
gts = (Xi-Xi0)*l;

% determint location of contact node in slip array
for j = 1:size(elec,1)
   if elec(i,1) == slip(j,1)
      gt_s = slip(j,2);
      count = j;
   end
end
t_tr = epsilon*(gts - gt_s);
```

```
% determine if slip occurs
if gts == 0
   nt = -1;
else
   nt = abs(gts-slip(i,2))/(gts-slip(i,2));
end

if abs(t_tr)-abs(Mu*epsilon*gns) <= 0
   Tt = t_tr;
   'stick';
else
   Tt = Mu*epsilon*abs(gns)*nt;
   slip2(count,2) = gt_s+1/epsilon*(abs(t_tr)-abs(Mu*epsilon*gns))*nt;
   'slip';
end

F_cie = epsilon*gns*l*Ns+Tt*l*(Ts+gns/l*N0s+gts/l*T0s)-
epsilon*gns*l*tan_beta*t1*(Ts+gns/l*N0s+gts/l*T0s);   % l should be
actual area of contact area within one element

a = ID(ID2(elec(i,1),1));
b = ID(ID2(elec(i,1),2));
c = ID(ID2(elec(i,2),1));
d = ID(ID2(elec(i,2),2));
e = ID(ID2(elec(i,2),3));
f = ID(ID2(elec(i,3),1));
g = ID(ID2(elec(i,3),2));
h = ID(ID2(elec(i,3),3));
j = ID(ID2(elec(i,4),1));
k = ID(ID2(elec(i,4),2));
l = ID(ID2(elec(i,5),1));
m = ID(ID2(elec(i,5),2));
temp = [a,b,c,d,e,f,g,h,j,k,l,m];
for n = 1:12
   F_ci(temp(n)) =F_ci(temp(n))+F_cie(n);
end
end
```

Assembly_ka.m

```
% tangent stiffness matrix from embed beam
```

```
function Ka =
Assembly_ka(x,y,Et,At,Eemb,Aemb,Iemb,Lt,Lemb,elet,eleemb,neq,ID,ID
2,ODB)
Ka = zeros(neq);              % Initialize global tangent stiffness matrix
x2 = ODB(1).Node.X;
y2 = ODB(1).Node.Y;

% Assemble embbeding beam element
for i = 1 : size((eleemb),1)
   xe = [x2(eleemb(i,1)),x2(eleemb(i,2))];
   ye = [y2(eleemb(i,1)),y2(eleemb(i,2))];
   if (xe(2) - xe(1)) >= 0;
      a = 2*pi+atan((ye(2)-ye(1))/(xe(2)-xe(1)));
   else
      a = pi+atan((ye(2)-ye(1))/(xe(2)-xe(1)));
   end
   ke = [Eemb*Aemb/Lemb(i),0,0,-
Eemb*Aemb/Lemb(i),0,0;0,12*Eemb*Iemb/Lemb(i)^3,6*Eemb*Iemb/Le
mb(i)^2,0,-
12*Eemb*Iemb/Lemb(i)^3,6*Eemb*Iemb/Lemb(i)^2;0,6*Eemb*Iemb/Le
mb(i)^2,4*Eemb*Iemb/Lemb(i),0,-
6*Eemb*Iemb/Lemb(i)^2,2*Eemb*Iemb/Lemb(i);-
Eemb*Aemb/Lemb(i),0,0,Eemb*Aemb/Lemb(i),0,0;0,-
12*Eemb*Iemb/Lemb(i)^3,-
6*Eemb*Iemb/Lemb(i)^2,0,12*Eemb*Iemb/Lemb(i)^3,-
6*Eemb*Iemb/Lemb(i)^2;0,6*Eemb*Iemb/Lemb(i)^2,2*Eemb*Iemb/Le
mb(i),0,-6*Eemb*Iemb/Lemb(i)^2,4*Eemb*Iemb/Lemb(i)];
   ke = TransformB(a)'*ke*TransformB(a);
   % Global element assembly for beam
   a = ID(ID2(eleemb(i,1),1));
   b = ID(ID2(eleemb(i,1),2));
   c = ID(ID2(eleemb(i,1),3));
   d = ID(ID2(eleemb(i,2),1));
   e = ID(ID2(eleemb(i,2),2));
   f = ID(ID2(eleemb(i,2),3));
   temp = [a,b,c,d,e,f];
   for n = 1:6
      for m = 1:6
         Ka(temp(n),temp(m)) = Ka(temp(n),temp(m))+ke(n,m);
      end
   end
```

```
end
```

Assembly_kac.m

```
% tangent stiffness matrix from anchorage constraint
function Kac = Assembly_kac(eleac,episilon,ID,ID2,neq)

Kac = zeros(neq);
% Assemble embbeding beam element
for i = 1 : size((eleac),1)
   ke = episilon*[1,0,-1,0,0;0,1,0,-1,0;-1,0,1,0,0;0,-1,0,1,0;0,0,0,0,0];
   % Global element assembly for beam
   a = ID(ID2(eleac(i,1),1));
   b = ID(ID2(eleac(i,1),2));
   c = ID(ID2(eleac(i,2),1));
   d = ID(ID2(eleac(i,2),2));
   e = ID(ID2(eleac(i,2),3));
   temp = [a,b,c,d,e];
   for n = 1:5
      for m = 1:5
         Kac(temp(n),temp(m)) = Kac(temp(n),temp(m))+ke(n,m);
      end
   end
end
```

Assembly_kb.m

```
% assemble nonlinear beam stiffness matrix kc
function Kb =
Assembly_kb(ID,ID2,neq,x,y,eleb,Lb,U,concrete_stress,concrete_strain,A
,S,I,hf,n,...
   steel_stress,steel_strain,zs,As,Ss,Is,Ec,Esteel)
Kb = zeros(neq);

for i = 1 :size(eleb,1)
   %transfor local coordinates to global coordinates
   xe = [x(eleb(i,1)),x(eleb(i,2))];
   ye = [y(eleb(i,1)),y(eleb(i,2))];
   if (xe(2) - xe(1)) >= 0;
      a = 2*pi+atan((ye(2)-ye(1))/(xe(2)-xe(1)));
   else
```

```
    a = pi+atan((ye(2)-ye(1))/(xe(2)-xe(1)));
  end

  % extract displacement vector for each nonlinear beam element
  Ue = TransformB(a)*[U(ID(ID2(eleb(i,1),1)));...
    U(ID(ID2(eleb(i,1),2)));...
    U(ID(ID2(eleb(i,1),3)));...
    U(ID(ID2(eleb(i,2),1)));...
    U(ID(ID2(eleb(i,2),2)));...
    U(ID(ID2(eleb(i,2),3)))];
  le = Lb(i);

  % compute B matrix for 6 Gauss pts within one element
  B = zeros(2,6,2);
  Xi = [-0.577350269,0.577350269]; % 2 pts Gauss quadrature
  W = [1,1]; % weights
  epsilon_e = zeros(2,2);
  for loop3 = 1:2
    B(:,:,loop3) = 1/le*[-1,0,0,1,0,0;0,6*Xi(loop3)/le,3*Xi(loop3)-1,0,-
6*Xi(loop3)/le,3*Xi(loop3)+1];
    epsilon_e(:,loop3) = B(:,:,loop3)*Ue;
  end

  epsilon_0 = epsilon_e(1,:)'*ones(1,n);  % axial strain on the ref axis
(constant through one element)
  culvature = epsilon_e(2,:);
  epsilon_0s = epsilon_e(1,:)'*ones(1,size(zs,2));
  epsilon_c = epsilon_0+[culvature(1)*hf;culvature(2)*hf];   % column
stands for each fiber section, row stands for each integration pts
  epsilon_s = epsilon_0s+[culvature(1)*zs;culvature(2)*zs];

  E = interp1(Ec(2,:),Ec(1,:),epsilon_c);
  Es = interp1(Esteel(2,:),Esteel(1,:),epsilon_s);

  % compute D & Dt for 6 Gauss pts within one element
  Dc = zeros(2,2,2);
  for loop1 = 1:2
    for loop2 = 1:n
      Dc(:,:,loop1) =
Dc(:,:,loop1)+E(loop1,loop2)*[A(loop2),S(loop2);S(loop2),I(loop2)];
    end
```

```
    end

    Ds = zeros(2,2,2);
    for loop1 = 1:2
        for loop2 = 1:size(zs,2)
            Ds(:,:,loop1) =
Ds(:,:,loop1)+Es(loop1,loop2)*[As(loop2),Ss(loop2);Ss(loop2),Is(loop2)];
        end
    end
    D = Dc+Ds;

    % compute element stiffness matrix and tangential matrix
    ke = zeros(6,6);

    for loop4 = 1:2
        ke = ke+le/2*W(loop4)*B(:,:,loop4)'*D(:,:,loop4)*B(:,:,loop4);
    end
    ke = TransformB(a)'*ke*TransformB(a);

    % Global element assembly for beam
    a = ID(ID2(eleb(i,1),1));
    b = ID(ID2(eleb(i,1),2));
    c = ID(ID2(eleb(i,1),3));
    d = ID(ID2(eleb(i,2),1));
    e = ID(ID2(eleb(i,2),2));
    f = ID(ID2(eleb(i,2),3));
    temp = [a,b,c,d,e,f];
    for loop10 = 1:6
        for loop11 = 1:6
            Kb(temp(loop10),temp(loop11)) =
Kb(temp(loop10),temp(loop11))+ke(loop10,loop11);
        end
    end
end
```

Assembly_kbkbt.m

```
% assemble nonlinear beam stiffness matrix kc
function [Kb,Kbt] =
Assembly_kbkbt(ID,ID2,neq,x,y,eleb,Lb,U,concrete_stress,concrete_strai
n,A,S,I,hf,n,...
```

```
   steel_stress,steel_strain,zs,As,Ss,Is)
Kb = zeros(neq);
Kbt = zeros(neq);

Xi = [-0.577350269,0.577350269]; % 2 pts Gauss quadrature
W = [1,1];                        % weights

for i = 1 :size(eleb,1)
   % extract displacement vector for each nonlinear beam element
   Ue = [U(ID(ID2(eleb(i,1),1)));...
      U(ID(ID2(eleb(i,1),2)));...
      U(ID(ID2(eleb(i,1),3)));...
      U(ID(ID2(eleb(i,2),1)));...
      U(ID(ID2(eleb(i,2),2)));...
      U(ID(ID2(eleb(i,2),3)))];
   le = Lb(i);

   % compute B matrix for 6 Gauss pts within one element
   B = zeros(2,6,2);
   epsilon_e = zeros(2,2);
   for loop3 = 1:2
      B(:,:,loop3) = 1/le*[-1,0,0,1,0,0;0,6*Xi(loop3)/le,3*Xi(loop3)-1,0,-
6*Xi(loop3)/le,3*Xi(loop3)+1];
      epsilon_e(:,loop3) = B(:,:,loop3)*Ue;
   end

   % axial strain on the ref axis (constant through one element)
   epsilon_0 = epsilon_e(1,:)'*ones(1,n);
   culvature = epsilon_e(2,:);
   epsilon_0s = epsilon_e(1,:)'*ones(1,size(zs,2));
   % column stands for each fiber section, row stands for each integration
pts
   epsilon_c = epsilon_0+[culvature(1)*hf;culvature(2)*hf];
   epsilon_s = epsilon_0s+[culvature(1)*zs;culvature(2)*zs];
   % approximate initial Young's modulus from user stress-strain data
   E0 = interp1(concrete_strain,concrete_stress,0.0001)/0.0001;
   E0s = interp1(steel_strain,steel_stress,0.0001)/0.0001;
   sigma_c = interp1(concrete_strain,concrete_stress,epsilon_c);
   sigma_s = interp1(steel_strain,steel_stress,epsilon_s);
   [E,Et] =
slopinterpolator(sigma_c,epsilon_c,concrete_stress,concrete_strain,E0);
```

```
    [Es,Ets] =
slopinterpolator(sigma_s,epsilon_s,steel_stress,steel_strain,E0s);

    % compute D & Dt for 6 Gauss pts within one element
    Dc = zeros(2,2,2);
    Dtc = zeros(2,2,2);

    for loop1 = 1:2
        for loop2 = 1:n
            Dc(:,:,loop1) =
Dc(:,:,loop1)+E(loop1,loop2)*[A(loop2),S(loop2);S(loop2),I(loop2)];
            Dtc(:,:,loop1) =
Dtc(:,:,loop1)+Et(loop1,loop2)*[A(loop2),S(loop2);S(loop2),I(loop2)];
        end
    end

    Ds = zeros(2,2,2);
    Dts = zeros(2,2,2);
    for loop1 = 1:2
        for loop2 = 1:size(zs,2)
            Ds(:,:,loop1) =
Ds(:,:,loop1)+Es(loop1,loop2)*[As(loop2),Ss(loop2);Ss(loop2),Is(loop2)];
            Dts(:,:,loop1) =
Dts(:,:,loop1)+Ets(loop1,loop2)*[As(loop2),Ss(loop2);Ss(loop2),Is(loop2)
];
        end
    end

    D = Dc+Ds;
    Dt = Dtc+Dts;

    % compute element stiffness matrix and tangential matrix
    ke = zeros(6,6);
    ket = zeros(6,6);
    for loop4 = 1:2
        ke = ke+le/2*W(loop4)*B(:,:,loop4)'*D(:,:,loop4)*B(:,:,loop4);
        ket = ket+le/2*W(loop4)*B(:,:,loop4)'*Dt(:,:,loop4)*B(:,:,loop4);
    end

    %transfor local coordinates to global coordinates
    xe = [x(eleb(i,1)),x(eleb(i,2))];
```

```
    ye = [y(eleb(i,1)),y(eleb(i,2))];
    if (xe(2) - xe(1)) >= 0;
        a = 2*pi+atan((ye(2)-ye(1))/(xe(2)-xe(1)));
    else
        a = pi+atan((ye(2)-ye(1))/(xe(2)-xe(1)));
    end
    ke = TransformB(a)'*ke*TransformB(a);
    ket = TransformB(a)'*ket*TransformB(a);
    % Global element assembly for beam

    a = ID(ID2(eleb(i,1),1));
    b = ID(ID2(eleb(i,1),2));
    c = ID(ID2(eleb(i,1),3));
    d = ID(ID2(eleb(i,2),1));
    e = ID(ID2(eleb(i,2),2));
    f = ID(ID2(eleb(i,2),3));
    temp = [a,b,c,d,e,f];
    for loop10 = 1:6
        for loop11 = 1:6
            Kb(temp(loop10),temp(loop11)) =
Kb(temp(loop10),temp(loop11))+ke(loop10,loop11);
            Kbt(temp(loop10),temp(loop11)) =
Kbt(temp(loop10),temp(loop11))+ket(loop10,loop11);
        end
    end
end
```

Assembly_kbt.m

```
% assemble nonlinear beam stiffness matrix kc
function Kbt =
Assembly_kbt(ID,ID2,neq,x,y,eleb,Lb,U,concrete_stress,concrete_strain,
A,S,I,hf,n,...
    steel_stress,steel_strain,zs,As,Ss,Is,Ect,Est)
Kbt = zeros(neq);

for i = 1 :size(eleb,1)
    %transfor local coordinates to global coordinates
    xe = [x(eleb(i,1)),x(eleb(i,2))];
    ye = [y(eleb(i,1)),y(eleb(i,2))];
```

```
if (xe(2) - xe(1)) >= 0;
   a = 2*pi+atan((ye(2)-ye(1))/(xe(2)-xe(1)));
else
   a = pi+atan((ye(2)-ye(1))/(xe(2)-xe(1)));
end

% extract displacement vector for each nonlinear beam element
Ue = TransformB(a)*[U(ID(ID2(eleb(i,1),1)));...
   U(ID(ID2(eleb(i,1),2)));...
   U(ID(ID2(eleb(i,1),3)));...
   U(ID(ID2(eleb(i,2),1)));...
   U(ID(ID2(eleb(i,2),2)));...
   U(ID(ID2(eleb(i,2),3)))];

le = Lb(i);

% compute B matrix for 6 Gauss pts within one element
B = zeros(2,6,2);
Xi = [-0.577350269,0.577350269];          % 2 pts Gauss quadrature
W = [1,1];                                % weights
epsilon_e = zeros(2,2);
for loop3 = 1:2
   B(:,:,loop3) = 1/le*[-1,0,0,1,0,0;0,6*Xi(loop3)/le,3*Xi(loop3)-1,0,-
6*Xi(loop3)/le,3*Xi(loop3)+1];
   epsilon_e(:,loop3) = B(:,:,loop3)*Ue;
end

% axial strain on the ref axis (constant through one element)
epsilon_0 = epsilon_e(1,:)'*ones(1,n);     culvature = epsilon_e(2,:);
epsilon_0s = epsilon_e(1,:)'*ones(1,size(zs,2));
% column stands for each fiber section, row stands for each integration
pts
epsilon_c = epsilon_0+[culvature(1)*hf;culvature(2)*hf];
epsilon_s = epsilon_0s+[culvature(1)*zs;culvature(2)*zs];
Et = interp1(Ect(2,:),Ect(1,:),epsilon_c);
Ets = interp1(Est(2,:),Est(1,:),epsilon_s);

% compute D & Dt for 6 Gauss pts within one element
Dtc = zeros(2,2,2);
for loop1 = 1:2
   for loop2 = 1:n
```

```
        Dtc(:,:,loop1) =
Dtc(:,:,loop1)+Et(loop1,loop2)*[A(loop2),S(loop2);S(loop2),I(loop2)];
      end
   end

   Dts = zeros(2,2,2);
   for loop1 = 1:2
      for loop2 = 1:size(zs,2)
         Dts(:,:,loop1) =
Dts(:,:,loop1)+Ets(loop1,loop2)*[As(loop2),Ss(loop2);Ss(loop2),Is(loop2)
];
      end
   end

   Dt = Dtc+Dts;

   % compute element stiffness matrix and tangential matrix
   ket = zeros(6,6);
   for loop4 = 1:2
      ket = ket+le/2*W(loop4)*B(:,:,loop4)'*Dt(:,:,loop4)*B(:,:,loop4);
   end
    ket = TransformB(a)'*ket*TransformB(a);

   % Global element assembly for beam
    a = ID(ID2(eleb(i,1),1));
   b = ID(ID2(eleb(i,1),2));
   c = ID(ID2(eleb(i,1),3));
   d = ID(ID2(eleb(i,2),1));
   e = ID(ID2(eleb(i,2),2));
   f = ID(ID2(eleb(i,2),3));
   temp = [a,b,c,d,e,f];
   for loop10 = 1:6
      for loop11 = 1:6
         Kbt(temp(loop10),temp(loop11)) =
Kbt(temp(loop10),temp(loop11))+ket(loop10,loop11);
      end
   end
end
```

Assembly_kns.m

```
% tangent stiffness matrix from normal contact
function Kns = Assembly_kns(epsilon,x,y,xi,yi,ID,ID2,elec,neq)
Kns = zeros(neq);  % Initialize contact element tangent stiffness matrix
Kns_num = zeros(neq);  % Initialize contact element tangent stiffness
matrix for numerically generated matrix
knse2 = zeros(12,12);   % Initialize numerically generated tangential
stiffness matrix
hi = 1e-15; %initilize increment to generate tangential stiffness matrix
for i = 1 : size((elec),1)
   %compute tangent stiffness matrix via complex step method
   for i2 =[1,2,3,4,6,7,9,10,11,12] %j loop to compute partial derivative
with respect to individual DOF of the contact element: master-x, master-y,
slave1-1x, slave1-y, slave1-r, slave2-x, slave2-y, slave2-r. slave1-r and
slave2-r are ZERO
      % current lengeth of one element
      x2=x;                %initilize x2 = x for numerical tangential
matrix computation later
      y2=y;                %initilize y2 = y for numerical tangential
matrix computation later
      switch i2
         case 1            %master-x
            x2(elec(i,1))=complex(x(elec(i,1)),hi);
         case 2            %master-y
            y2(elec(i,1))=complex(y(elec(i,1)),hi);
         case 3            %slave1-x
            x2(elec(i,2))=complex(x(elec(i,2)),hi);
         case 4            %slave1-y
            y2(elec(i,2))=complex(y(elec(i,2)),hi);
         case 6            %slave2-x
            x2(elec(i,3))=complex(x(elec(i,3)),hi);
         case 7            %slave2-y
            y2(elec(i,3))=complex(y(elec(i,3)),hi);
         case 9            %truss_left-x
            x2(elec(i,4))=complex(x(elec(i,4)),hi);
         case 10           %truss_left-y
            y2(elec(i,4))=complex(y(elec(i,4)),hi);
         case 11           %truss_right-x
            x2(elec(i,5))=complex(x(elec(i,5)),hi);
         case 12           %truss_right-y
            y2(elec(i,5))=complex(y(elec(i,5)),hi);
      end
```

```
l = sqrt((y2(elec(i,3))-y2(elec(i,2)))^2+(x2(elec(i,3))-x2(elec(i,2)))^2);

% unit tangent vector
t(1) = (x2(elec(i,3))-x2(elec(i,2)))/l;
t(2) = (y2(elec(i,3))-y2(elec(i,2)))/l;

% unit normal vector
n(1) = -t(2);
n(2) = t(1);

% unit middle vector between two truss element where contact
master
% node is in the middle
n1(1) = (xi(elec(i,4))-xi(elec(i,1)))/sqrt((xi(elec(i,4))-
xi(elec(i,1)))^2+(yi(elec(i,4))-yi(elec(i,1)))^2)/2+(xi(elec(i,5))-
xi(elec(i,1)))/sqrt((xi(elec(i,5))-xi(elec(i,1)))^2+(yi(elec(i,5))-
yi(elec(i,1)))^2)/2;
n1(2) = (yi(elec(i,4))-yi(elec(i,1)))/sqrt((xi(elec(i,4))-
xi(elec(i,1)))^2+(yi(elec(i,4))-yi(elec(i,1)))^2)/2+(yi(elec(i,5))-
yi(elec(i,1)))/sqrt((xi(elec(i,5))-xi(elec(i,1)))^2+(yi(elec(i,5))-
yi(elec(i,1)))^2)/2;

%if tendon is complete straight, n1 vector will become [0,0]. In
%that case, n1 will be corrected as a vector orthgonal to either of
%the truss element vectors
if n1 == [0,0]
  n1(1)=-(yi(elec(i,4))-yi(elec(i,1)));
  n1(2)=xi(elec(i,4))-xi(elec(i,1));
end
n1=n1/norm(n1,2);
tan_beta = abs(n1(1)*n(2)-n1(2)*n(1))/(n1(1)*n(1)+n1(2)*n(2));

%compute correction force unit directional vector t1
if (abs( n1(1)*t(1)+n1(2)*t(2))) == 0
  t1 = 0;
else
  t1 = (n1(1)*t(1)+n1(2)*t(2))/(abs( n1(1)*t(1)+n1(2)*t(2)));
end

Xi = 1/l*[x2(elec(i,1))-x2(elec(i,2)),y2(elec(i,1))-y2(elec(i,2))]*t';
Xi0 = 1/l*[xi(elec(i,1))-xi(elec(i,2)),yi(elec(i,1))-yi(elec(i,2))]*t';
```

```
        Ns = [n,-(1-Xi)*n,0,-Xi*n,0,0,0,0,0]';
        N0s = [0,0,-n,0,n,0,0,0,0,0]';
        Ts = [t,-(1-Xi)*t,0,-Xi*t,0,0,0,0,0]';
        T0s = [0,0,-t,0,t,0,0,0,0,0]';
        gns = dot([x2(elec(i,1));y2(elec(i,1))]-(1-
Xi)*[x2(elec(i,2));y2(elec(i,2))]-Xi*[x2(elec(i,3));y2(elec(i,3))],-n);
        gts = (Xi-Xi0)*l;
        % l should be actual area of contact area within one element
        Gn = epsilon*gns*l*Ns-
epsilon*gns*l*tan_beta*t1*(Ts+gns/l*N0s+gts/l*T0s);
        knse2(1,i2) = imag(Gn(1)/hi);
        knse2(2,i2) = imag(Gn(2)/hi);
        knse2(3,i2) = imag(Gn(3)/hi);
        knse2(4,i2) = imag(Gn(4)/hi);
        knse2(5,i2) = 0;
        knse2(6,i2) = imag(Gn(6)/hi);
        knse2(7,i2) = imag(Gn(7)/hi);
        knse2(8,i2) = 0;
        knse2(9,i2) = imag(Gn(9)/hi);
        knse2(10,i2) = imag(Gn(10)/hi);
        knse2(11,i2) = imag(Gn(11)/hi);
        knse2(12,i2) = imag(Gn(12)/hi);
    end

    % Global element assembly
    a = ID(ID2(elec(i,1),1));
    b = ID(ID2(elec(i,1),2));
    c = ID(ID2(elec(i,2),1));
    d = ID(ID2(elec(i,2),2));
    e = ID(ID2(elec(i,2),3));
    f = ID(ID2(elec(i,3),1));
    g = ID(ID2(elec(i,3),2));
    h = ID(ID2(elec(i,3),3));
    j = ID(ID2(elec(i,4),1));
    k = ID(ID2(elec(i,4),2));
    l = ID(ID2(elec(i,5),1));
    m = ID(ID2(elec(i,5),2));
    temp = [a,b,c,d,e,f,g,h,j,k,l,m];
    for n = 1:12
        for m = 1:12
            Kns(temp(n),temp(m)) = Kns(temp(n),temp(m))+knse2(n,m);
```

```
      end
   end
end
```

Assembly_kps.m

```
% tangent stiffness maxtrix from prestress tendon
function Kps = Assembly_kps(x,y,At,Lt,elet,neq,ID,ID2,Eps,U,ODB)
Kps = zeros(neq);   % Initialize global tangent stiffness matrix

x2 = ODB(1).Node.X;
y2 = ODB(1).Node.Y;

% Assemble truss element
for i = 1 : size((elet),1)
   xe = [x(elet(i,1)),x(elet(i,2))];
   ye = [y(elet(i,1)),y(elet(i,2))];
   if (xe(2) - xe(1)) >= 0;
      a = 2*pi+atan((ye(2)-ye(1))/(xe(2)-xe(1)));
   else
      a = pi+atan((ye(2)-ye(1))/(xe(2)-xe(1)));
   end

      % extract displacement vector for each nonlinear beam element
   Ue = TransformT(a)*[U(ID(ID2(elet(i,1),1)));...
      U(ID(ID2(elet(i,1),2)));...
      U(ID(ID2(elet(i,2),1)));...
      U(ID(ID2(elet(i,2),2)))];
   e_truss = 1/Lt(i)*[-1,0,1,0]*Ue;

   Et = interp1(Eps(2,:),Eps(1,:),e_truss);

   if ID2(elet(i,1),3) > 0 || ID2(elet(i,2),3) > 0
      ke = [Et*At/Lt(i),0,-Et*At/Lt(i),0,0;0,0,0,0,0;-
Et*At/Lt(i),0,Et*At/Lt(i),0,0;0,0,0,0,0;0,0,0,0,0];
      ke = TransformTB(a)'*ke*TransformTB(a);
      % Global element assembly for truss-beam connection
      if ID2(elet(i,1),3) == 0
         a = ID(ID2(elet(i,1),1));
         b = ID(ID2(elet(i,1),2));
         c = ID(ID2(elet(i,2),1));
```

```
        d = ID(ID2(elet(i,2),2));
        e = ID(ID2(elet(i,2),3));
      else
        a = ID(ID2(elet(i,2),1));
        b = ID(ID2(elet(i,2),2));
        c = ID(ID2(elet(i,1),1));
        d = ID(ID2(elet(i,1),2));
        e = ID(ID2(elet(i,1),3));
      end
      temp = [a,b,c,d,e];
      for n = 1:5
        for m = 1:5
          Kps(temp(n),temp(m)) = Kps(temp(n),temp(m))+ke(n,m);
        end
      end
    else
      ke = [Et*At/Lt(i),0,-Et*At/Lt(i),0;0,0,0,0;-
Et*At/Lt(i),0,Et*At/Lt(i),0;0,0,0,0];
      ke = TransformT(a)'*ke*TransformT(a);
      % Global element assembly for truss
      a = ID(ID2(elet(i,1),1));
      b = ID(ID2(elet(i,1),2));
      c = ID(ID2(elet(i,2),1));
      d = ID(ID2(elet(i,2),2));
      temp = [a,b,c,d];
      for n = 1:4
        for m = 1:4
          Kps(temp(n),temp(m)) = Kps(temp(n),temp(m))+ke(n,m);
        end
      end
    end
end
```

Assembly_kpst.m

```
% tangent stiffness maxtrix from prestress tendon
function Kpst = Assembly_kpst(x,y,At,Lt,elet,neq,ID,ID2,Epst)
% Initialize global tangent stiffness matrix
Kpst = zeros(neq);

% Assemble truss element
```

```
for i = 1 : size((elet),1)
   xe = [x(elet(i,1)),x(elet(i,2))];
   ye = [y(elet(i,1)),y(elet(i,2))];
   if (xe(2) - xe(1)) >= 0;
      a = 2*pi+atan((ye(2)-ye(1))/(xe(2)-xe(1)));
   else
      a = pi+atan((ye(2)-ye(1))/(xe(2)-xe(1)));
   end

   current_leng = sqrt((ye(2)-ye(1))^2+(xe(2)-xe(1))^2);
   e_truss = (current_leng-Lt(i))/Lt(i);
   Et = interp1(Epst(2,:),Epst(1,:),e_truss);

   if ID2(elet(i,1),3) > 0 || ID2(elet(i,2),3) > 0
      ke = [Et*At/Lt(i),0,-Et*At/Lt(i),0,0;0,0,0,0,0;-
Et*At/Lt(i),0,Et*At/Lt(i),0,0;0,0,0,0,0;0,0,0,0,0];
      ke = TransformTB(a)'*ke*TransformTB(a);

      % Global element assembly for truss-beam connection
      if ID2(elet(i,1),3) == 0
         a = ID(ID2(elet(i,1),1));
         b = ID(ID2(elet(i,1),2));
         c = ID(ID2(elet(i,2),1));
         d = ID(ID2(elet(i,2),2));
         e = ID(ID2(elet(i,2),3));
      else
         a = ID(ID2(elet(i,2),1));
         b = ID(ID2(elet(i,2),2));
         c = ID(ID2(elet(i,1),1));
         d = ID(ID2(elet(i,1),2));
         e = ID(ID2(elet(i,1),3));
      end
      temp = [a,b,c,d,e];
      for n = 1:5
         for m = 1:5
            Kpst(temp(n),temp(m)) = Kpst(temp(n),temp(m))+ke(n,m);
         end
      end
   else
      ke = [Et*At/Lt(i),0,-Et*At/Lt(i),0;0,0,0,0;-
Et*At/Lt(i),0,Et*At/Lt(i),0;0,0,0,0];
```

```
    ke = TransformT(a)'*ke*TransformT(a);

    % Global element assembly for truss
    a = ID(ID2(elet(i,1),1));
    b = ID(ID2(elet(i,1),2));
    c = ID(ID2(elet(i,2),1));
    d = ID(ID2(elet(i,2),2));
    temp = [a,b,c,d];
    for n = 1:4
      for m = 1:4
        Kpst(temp(n),temp(m)) = Kpst(temp(n),temp(m))+ke(n,m);
      end
    end
  end
end
```

Assembly_kts.m

```
% tangent stiffness matrix from tangential contact
function Kts = Assembly_kts(epsilon,x,y,xi,yi,ID,ID2,elec,neq,Mu,slip)

Kts = zeros(neq);  % Initialize contact element tangent stiffness matrix
for i = 1 : size((elec),1)

  % current length of one element
  l = sqrt((y(elec(i,3))-y(elec(i,2)))^2+(x(elec(i,3))-x(elec(i,2)))^2);

  % unit tangent vector
  t(1) = (x(elec(i,3))-x(elec(i,2)))/l;
  t(2) = (y(elec(i,3))-y(elec(i,2)))/l;

  % unit normal vector
  n(1) = -t(2);
  n(2) = t(1);

  Xi = 1/l*[x(elec(i,1))-x(elec(i,2)),y(elec(i,1))-y(elec(i,2))]*t';
  Xi0 = 1/l*[xi(elec(i,1))-xi(elec(i,2)),yi(elec(i,1))-yi(elec(i,2))]*t';
  Ns = [n,-(1-Xi)*n,0,-Xi*n,0]';
  N0s = [0,0,-n,0,n,0]';
  Ts = [t,-(1-Xi)*t,0,-Xi*t,0]';
  T0s = [0,0,-t,0,t,0]';
```

```
   temp = [x(elec(i,1));y(elec(i,1))]-(1-Xi)*[x(elec(i,2));y(elec(i,2))]-
Xi*[x(elec(i,3));y(elec(i,3))];
   if temp(1)/t(2) < 0
      gns = -norm(temp,2);
   else
      gns = norm(temp,2);
   end

   if gns == 0
      ns = 1;
   else
      ns = abs(gns)/gns;
   end

   gts = (Xi-Xi0)*l;

   Tss = (Ts+gns/l*N0s+gts/l*T0s);
   if gts == 0
      nt = 1;
   else
      nt = -abs(gts-slip(i,2))/(gts-slip(i,2));
   end

   ktse = nt*Mu*epsilon*l*(ns*Tss*Ns'+abs(gns)/l*(N0s*Ns'+T0s*Tss'));

% Global element assembly
   a = ID(ID2(elec(i,1),1));
   b = ID(ID2(elec(i,1),2));
   c = ID(ID2(elec(i,2),1));
   d = ID(ID2(elec(i,2),2));
   e = ID(ID2(elec(i,2),3));
   f = ID(ID2(elec(i,3),1));
   g = ID(ID2(elec(i,3),2));
   h = ID(ID2(elec(i,3),3));
   temp = [a,b,c,d,e,f,g,h];
   for n = 1:8
      for m = 1:8
         Kts(temp(n),temp(m)) = Kts(temp(n),temp(m))+ktse(n,m);
      end
   end
```

end

Assembly_kts2.m

```
% tangent stiffness matrix from tangential contact (will test stik/slip
condition)
function Kts =
Assembly_kts2(epsilon,x,y,xi,yi,ID,ID2,elec,neq,Mu,slip,slip2,count2)

Kts = zeros(neq); % Initialize contact element tangent stiffness matrix
Kts_num = zeros(neq);  % Initialize contact element tangent stiffness
matrix for numerically generated matrix
ktse2 = zeros(8,8);          % Initialize numerically generated tangential
stiffness matrix
hi = 1e-15;                  %initilize increment to generate tangential
stiffness matrix
%initilize x2 = x for numerical tangential matrix computation later
x2=x;
%initilize y2 = y for numerical tangential matrix computation later
y2=y;
for i = 1 : size((elec),1)
   %compute tangent stiffness matrix via complex step method
   ktse2 = zeros(8,8);        % Initialize numerically generated tangential
stiffness matrix
   for i2 =[1,2,3,4,6,7]      %j loop to compute partial derivative with
respect to individual DOF of the contact element: master-x, master-y,
slave1-1x, slave1-y, slave1-r, slave2-x, slave2-y, slave2-r. slave1-r and
slave2-r are ZERO
       x2(elec(i,1))=complex(x(elec(i,1)),hi);
       y2(elec(i,1))=complex(y(elec(i,1)),hi);
       x2(elec(i,2))=complex(x(elec(i,2)),hi);
       y2(elec(i,2))=complex(y(elec(i,2)),hi);
       x2(elec(i,3))=complex(x(elec(i,3)),hi);
       y2(elec(i,3))=complex(y(elec(i,3)),hi);
       l = sqrt((y2(elec(i,3))-y2(elec(i,2)))^2+(x2(elec(i,3))-x2(elec(i,2)))^2);

       % unit tangent vector
       t(1) = (x2(elec(i,3))-x2(elec(i,2)))/l;
       t(2) = (y2(elec(i,3))-y2(elec(i,2)))/l;

       % unit normal vector
```

```
    n(1) = -t(2);
    n(2) = t(1);

    Xi = 1/l*[x2(elec(i,1))-x2(elec(i,2)),y2(elec(i,1))-y2(elec(i,2))]*t';
    Xi0 = 1/l*[xi(elec(i,1))-xi(elec(i,2)),yi(elec(i,1))-yi(elec(i,2))]*t';
    Ns = [n,-(1-Xi)*n,0,-Xi*n,0]';
    N0s = [0,0,-n,0,n,0]';
    Ts = [t,-(1-Xi)*t,0,-Xi*t,0]';
    T0s = [0,0,-t,0,t,0]';

    gns = dot([x2(elec(i,1));y2(elec(i,1))]-(1-
Xi)*[x2(elec(i,2));y2(elec(i,2))]-Xi*[x2(elec(i,3));y2(elec(i,3))],n);
    gts = (Xi-Xi0)*l;

    % determint location of contact node in slip array
    for j = 1:size(elec,1)
      if elec(i,1) == slip(j,1)
        gt_s = slip(j,2);
        count = j;
      end
    end

    t_tr = epsilon*(gts - gt_s);

    % determine if slip occurs
    if gts == 0
      nt = -1;
    else
      nt = abs(gts-slip(i,2))/(gts-slip(i,2));
    end

    if abs(t_tr)-abs(Mu*epsilon*gns) <= 0
      Tt = t_tr;
      'stick';
    else
      Tt = Mu*epsilon*abs(gns)*nt;
      slip2(count,2) = gt_s+1/epsilon*(abs(t_tr)-
abs(Mu*epsilon*gns))*nt;
      'slip';
    end
    if Mu == 0
```

```
            % frictionless contact
          Gt = zeros(8);
       Else
            % l should be actual area of contact area within one element
          Gt = 1t*l*(Ts+gns/l*N0s+gts/l*T0s);
       end
       ktse2(1,i2) = imag(Gt(1)/hi);
       ktse2(2,i2) = imag(Gt(2)/hi);
       ktse2(3,i2) = imag(Gt(3)/hi);
       ktse2(4,i2) = imag(Gt(4)/hi);
       ktse2(5,i2) = 0;
       ktse2(6,i2) = imag(Gt(6)/hi);
       ktse2(7,i2) = imag(Gt(7)/hi);
       ktse2(8,i2) = 0;
    end

    % Global element assembly
    a = ID(ID2(elec(i,1),1));
    b = ID(ID2(elec(i,1),2));
    c = ID(ID2(elec(i,2),1));
    d = ID(ID2(elec(i,2),2));
    e = ID(ID2(elec(i,2),3));
    f = ID(ID2(elec(i,3),1));
    g = ID(ID2(elec(i,3),2));
    h = ID(ID2(elec(i,3),3));
    temp = [a,b,c,d,e,f,g,h];
    for n = 1:8
       for m = 1:8
          Kts(temp(n),temp(m)) = Kts(temp(n),temp(m))+ktse2(n,m);
       end
    end
end
```

Assembly_kts3.m

```
% tangent stiffness matrix from tangential contact (stick behavior only)
function Kts = Assembly_kts3(epsilon,x,y,xi,yi,ID,ID2,elec,neq,Mu,slip)

% Initialize contact element tangent stiffness matrix
Kts = zeros(neq);
```

```
for i = 1 : size((elec),1)
   % current lengeth of one element
   l = sqrt((y(elec(i,3))-y(elec(i,2)))^2+(x(elec(i,3))-x(elec(i,2)))^2);

   % unit tangent vector
   t(1) = (x(elec(i,3))-x(elec(i,2)))/l;
   t(2) = (y(elec(i,3))-y(elec(i,2)))/l;

   % unit normal vector
   n(1) = -t(2);
   n(2) = t(1);

   Xi = 1/l*[x(elec(i,1))-x(elec(i,2)),y(elec(i,1))-y(elec(i,2))]*t';
   Xi0 = 1/l*[xi(elec(i,1))-xi(elec(i,2)),yi(elec(i,1))-yi(elec(i,2))]*t';
   Ns = [n,-(1-Xi)*n,0,-Xi*n,0]';
   N0s = [0,0,-n,0,n,0]';
   Ts = [t,-(1-Xi)*t,0,-Xi*t,0]';
   T0s = [0,0,-t,0,t,0]';

   temp = [x(elec(i,1));y(elec(i,1))]-(1-Xi)*[x(elec(i,2));y(elec(i,2))]-
Xi*[x(elec(i,3));y(elec(i,3))];
   if temp(1)/t(2) < 0
      gns = -norm(temp,2);
   else
      gns = norm(temp,2);
   end

   if gns == 0
      ns = 1;
   else
      ns = abs(gns)/gns;
   end

   gts = (Xi-Xi0)*l;

   Tss = (Ts+gns/l*N0s+gts/l*T0s);
   if gts == 0
      nt = 1;
   else
      nt = -abs(gts-slip(i,2))/(gts-slip(i,2));
   end
```

```
    for j = 1:size(elec,1)
        if elec(i,1) == slip(j,1)
            gt_s = slip(j,2);
            count = j;
        end
    end

    t_tr = epsilon*(gts - gt_s);
    ktse = epsilon*l*(Tss*Tss');

    % Global element assembly
    a = ID(ID2(elec(i,1),1));
    b = ID(ID2(elec(i,1),2));
    c = ID(ID2(elec(i,2),1));
    d = ID(ID2(elec(i,2),2));
    e = ID(ID2(elec(i,2),3));
    f = ID(ID2(elec(i,3),1));
    g = ID(ID2(elec(i,3),2));
    h = ID(ID2(elec(i,3),3));
    temp = [a,b,c,d,e,f,g,h];
    for n = 1:8
        for m = 1:8
            Kts(temp(n),temp(m)) = Kts(temp(n),temp(m))+ktse(n,m);
        end
    end
end
```

Disp_residual.m

```
% residual measure v0.1
function normresidual =
Disp_residual(U,d_F,alpha,nd,neq,xi,yi,ODB,step,...

epsilon,ID,ID2,elec,slip,Mu,Ka,F_jacking1,nnd,residual_Fi,epsilon2,anch
or_u,anchor_v,eleac)

% concrete unaxial stress-strain
concrete_stress = ODB(step).Material.Concrete(:,1)';
concrete_strain = ODB(step).Material.Concrete(:,2)';
% extend stress-strain curve
```

```
concrete_stress =
[concrete_stress(1),concrete_stress,concrete_stress(size(concrete_stress,2)
)];
concrete_strain =
[10*concrete_strain(1),concrete_strain,10*concrete_strain(size(concrete_s
train,2))];

% concrete beam section geometry
n_fiber = ODB(step).BeamElement.n;
hf = ODB(step).BeamElement.hf;
A = ODB(step).BeamElement.A;
S = ODB(step).BeamElement.S;
I = ODB(step).BeamElement.I;

% non-prestressed steel unaxial stress-strain
steel_stress = ODB(step).Material.Steel(:,1)';
steel_strain = ODB(step).Material.Steel(:,2)';
% extend stress-strain curve
steel_stress =
[steel_stress(1),steel_stress,steel_stress(size(steel_stress,2))];
steel_strain =
[10*steel_strain(1),steel_strain,10*steel_strain(size(steel_strain,2))];

% non-prestressed steel geometry
zs = ODB(step).BeamElement.zs;
As = ODB(step).BeamElement.As;
Ss = ODB(step).BeamElement.Ss;
Is = ODB(step).BeamElement.Is;

Lb = ODB(step).BeamElement.Length;
eleb = ODB(step).BeamElement.Nodes;

Ec = ODB(step).Material.Ec;
Es = ODB(step).Material.Es;

% Determine if it is the first loop for displacement increase vector
d_E = zeros(nd,1);

% Calculate residual force vector f_F
% Calculate current displacement U = U+d_F
U2 = U+[d_E;alpha*d_F];
```

```
% Update current node position
for i = 1:nnd
   x2(i) = xi(i)+U2(ID(ID2(i,1)));
   y2(i) = yi(i)+U2(ID(ID2(i,2)));
end

% Update residual force vector from anchorage contribution
if ODB(step).anchor == 0
   F_ac = zeros(neq,1);
else
   % Update residual force vector from anchorage contribution
   F_ac =
Assembly_F_aci(epsilon2,x2,y2,anchor_u,anchor_v,ID,ID2,eleac,neq);
end

[F_c,ODB(step).slip] =
Assembly_F_ci(epsilon,x2,y2,xi,yi,ID,ID2,elec,neq,slip,U2,Mu);

% Update stiffness maxtrix from nonlienar beam
Kb =
Assembly_kb(ID,ID2,neq,x2,y2,eleb,Lb,U2,concrete_stress,concrete_strai
n,A,S,I,hf,n_fiber,...
   steel_stress,steel_strain,zs,As,Ss,Is,Ec,Es);

% Update residual force vector from non-contact contribution
F_nc = (Ka+Kb)*U2;
F = -F_ac+F_c+F_nc;

normresidual = norm(-F(nd+1:neq)+residual_Fi+F_jacking1(nd+1:neq));
```

ElementLength.m

```
% Calculate element length
function L = ElementLength(ElementNodes,NodeX,NodeY) % Node
Label, X coordinates, Y coordinates)
for i = 1:size(ElementNodes,1)
   L(i) = sqrt((NodeX(ElementNodes(i,1))-
NodeX(ElementNodes(i,2)))^2+(NodeY(ElementNodes(i,1))-
NodeY(ElementNodes(i,2)))^2);
end
```

ID_array.m

```
% Genereate ID array
function [ID,d] = ID_array(nd,neq,flags,e_bc)
count = 0;
count1 = 0;
for i = 1:neq
  if flags(i) == 2              % check if essential boundary
    count = count + 1;
    ID(i) = count;              % number first the degrees-of-freedom on
essential boundary
    d(count)= e_bc(i);          % store the reordered values of essential B.C
  else
    count1 = count1 + 1;
    ID(i) = nd + count1;
  end
end
d = d';
```

ID2_array.m

```
% Genereate ID2 array
function ID2 = ID2_array(nnd,nddof)
ID2 = zeros(nnd,3);
count = 0;
for i = 1:nnd
   if nddof(i) == 2              % check if node on slide line
      for j = 1:2
         ID2(i,j) = count+1;
         count = count+1;
      end
   else
      for j = 1:3
         ID2(i,j) = count+1;
         count = count+1;
      end
   end
end
```

InitialCondition.m

```
% Generate initial condition for residual force and displacement vector
function newvector = InitialCondition(U,V,T,boundary)
count = 0;
```

```
oldvector = [U;V;T];
newvector = zeros(size(boundary,1)*3-sum(boundary(:) == 0),1);  %
Prelocate newvector to increase speed
for i = 1:size(boundary,1)
   for j = 1:3
      if boundary(i,j) > 0;
         count = count+1;
         newvector(count) = oldvector(j,i);
      end
   end
end
```

LineSearch.m

```
% Line search (minimize residual_f by modify displacement increment
vector)
function df_m = LineSearch(U,d_F,nd,neq,xi,yi,ODB,step,...

epsilon,ID,ID2,elec,slip,Mu,Ka,F_jacking1,nnd,residual_Fi,epsilon2,anch
or_u,anchor_v,eleac)
ctrl = 0.1;
df_m = d_F;
flag = 2;

while flag > 1
   Ftest_l = Disp_residual(U,df_m,1-ctrl,nd,neq,xi,yi,ODB,step,...

epsilon,ID,ID2,elec,slip,Mu,Ka,F_jacking1,nnd,residual_Fi,epsilon2,anch
or_u,anchor_v,eleac);
   Ftest_m = Disp_residual(U,df_m,1,nd,neq,xi,yi,ODB,step,...

epsilon,ID,ID2,elec,slip,Mu,Ka,F_jacking1,nnd,residual_Fi,epsilon2,anch
or_u,anchor_v,eleac);
   Ftest_r = Disp_residual(U,df_m,1+ctrl,nd,neq,xi,yi,ODB,step,...

epsilon,ID,ID2,elec,slip,Mu,Ka,F_jacking1,nnd,residual_Fi,epsilon2,anch
or_u,anchor_v,eleac);
   if Ftest_l > Ftest_m && Ftest_m > Ftest_r
      df_l = df_m;
      count1 = 2;
      while Ftest_l > Ftest_m
```

```
        df_m = count1*df_m;
        count1 = count1+1;
        Ftest_l = Disp_residual(U,df_m,1-ctrl,nd,neq,xi,yi,ODB,step,...

epsilon,ID,ID2,elec,slip,Mu,Ka,F_jacking1,nnd,residual_Fi,epsilon2,anch
or_u,anchor_v,eleac);
        Ftest_m = Disp_residual(U,df_m,1,nd,neq,xi,yi,ODB,step,...

epsilon,ID,ID2,elec,slip,Mu,Ka,F_jacking1,nnd,residual_Fi,epsilon2,anch
or_u,anchor_v,eleac);
      end
      df_r = df_m;
      df_m = 0.5*(df_l+df_r);
      flag = 2;

   elseif Ftest_l < Ftest_m && Ftest_m < Ftest_r
      df_r = df_m;
      df_l = 0*df_m;
      df_m = 0.5*(df_l+df_r);
      flag = 3;

   elseif Ftest_l < Ftest_m && Ftest_m > Ftest_r
      if flag == 2
         df_r = df_m;
         df_m = 0.5*(df_l+df_r);
      elseif flag == 3
         df_l = df_m;
         df_m = 0.5*(df_l+df_r);
      end

   elseif Ftest_l > Ftest_m && Ftest_m < Ftest_r
      flag = 1;
   end
end
```

Solver for frictionless contact (perfectly unbonded condition)
MainSolver3.m

```
% Solver main function for frictionless contact
function ODB = MainSolver3(ODB,step)
```

```
xi = ODB(step).Node.X;
yi = ODB(step).Node.Y;
Et = ODB(step).TrussElement.E(1);
At = ODB(step).TrussElement.A(1);
Eemb = ODB(step).EmbeddingElement.E(1);
Aemb = ODB(step).EmbeddingElement.A(1);
Iemb = ODB(step).EmbeddingElement.I(1);
Lt = ODB(step).TrussElement.Length;
Lb = ODB(step).BeamElement.Length;
Lemb = ODB(step).EmbeddingElement.Length;
elet = ODB(step).TrussElement.Nodes;
eleb = ODB(step).BeamElement.Nodes;
eleac = ODB(step).AnchorElement.Nodes;
elec = ODB(step).ContactElement.Nodes;
eleemb = ODB(step).EmbeddingElement.Nodes;
neq = ODB(step).neq;
nd = ODB(step).nd;
nnd = ODB(step).nnd;
ID = ODB(step).ID;
d_E = ODB(step).d_E;
ID2 = ODB(step).ID2;
slip = ODB(step).slip;
Mu = ODB(step).Material.Mu;
Uo = InitialCondition(ODB(step).Node.U,ODB(step).Node.V,...
   ODB(step).Node.T,ODB(step).Node.Boundary);
residual_Fo = InitialCondition(ODB(step).Node.Fx_residual,...

ODB(step).Node.Fy_residual,ODB(step).Node.M_residual,ODB(step).No
de.Boundary);
epsilon = ODB(step).ContactElement.Epsilon;
epsilon2 = ODB(step).AnchorElement.Epsilon;
x = xi;
y = yi;

% concrete unaxial stress-strain
concrete_stress = ODB(step).Material.Concrete(:,1)';
concrete_strain = ODB(step).Material.Concrete(:,2)';
% concrete beam section geometry
n_fiber = ODB(step).BeamElement.n;
hf = ODB(step).BeamElement.hf;
A = ODB(step).BeamElement.A;
```

```
S = ODB(step).BeamElement.S;
I = ODB(step).BeamElement.I;

% non-prestressed steel unaxial stress-strain
steel_stress = ODB(step).Material.Steel(:,1)';
steel_strain = ODB(step).Material.Steel(:,2)';

Ec = ODB(step).Material.Ec;
Es = ODB(step).Material.Es;
Eps = ODB(step).Material.Eps;
Ect = ODB(step).Material.Ect;
Est = ODB(step).Material.Est;
Epst = ODB(step).Material.Epst;

% non-prestressed steel geometry
zs = ODB(step).BeamElement.zs;
As = ODB(step).BeamElement.As;
Ss = ODB(step).BeamElement.Ss;
Is = ODB(step).BeamElement.Is;

% Determine if at anchorage stage and caluclate relative disp between two
% nodes before prestressing
if ODB(step).anchor == 1 && step == 2
  for i = 1:size(eleac,1)
    anchor_u(i) = x(eleac(i,2))+Uo(ID2(eleac(i,2),1))...
      -x(eleac(i,1))-Uo(ID2(eleac(i,1),1))
    anchor_v(i) = y(eleac(i,2))+Uo(ID2(eleac(i,2),2))...
      -y(eleac(i,1))-Uo(ID2(eleac(i,1),2))
  end
elseif ODB(step).anchor == 0
  anchor_u = zeros(size(eleac,1),1);
  anchor_v = zeros(size(eleac,1),1);
elseif step >= 3
  anchor_u = ODB(2).anchor_u;
  anchor_v = ODB(2).anchor_v;
end
ODB(step).anchor_u = anchor_u;
ODB(step).anchor_v = anchor_v;

% Transfer Uo residual_Fo from origional node ordering to new ordering
U = Reorder(ID,Uo);
```

```
residual_F = Reorder(ID,residual_Fo);
residual_F = residual_F(nd+1:neq);
residual_Fi = residual_F;

% tangent stiffness matrix from anchorage
if ODB(step).anchor == 0
   Kac = zeros(neq);
else
   Kac = Assembly_kac(eleac,epsilon,ID,ID2,neq);
end

% tangent stiffness matrix from normal contact
Kns = Assembly_kns(epsilon,x,y,xi,yi,ID,ID2,elec,neq);
% tangent stiffness matrix from tagent contact (slip)
Kts = Assembly_kts2(epsilon,x,y,xi,yi,ID,ID2,elec,neq,Mu,slip);

% Tangent stiffness matrix from linear elements (constant considering
elasticity and linear geometry)
Ka =
Assembly_ka(xi,yi,Et,At,Eemb,Aemb,Iemb,Lt,Lemb,elet,eleemb,neq,ID,I
D2,ODB);

% Tanget stiffness matrix from ptrestressed steel truss element
Kpst = Assembly_kpst(xi,yi,At,Lt,elet,neq,ID,ID2,Epst,U);

% Tangent stiffness matrix from nonlinear beam
Kbt =
Assembly_kbt(ID,ID2,neq,xi,yi,eleb,Lb,U,concrete_stress,concrete_strain,
A,S,I,hf,n_fiber,...
   steel_stress,steel_strain,zs,As,Ss,Is,Ect,Est);

% Sum up tagent stiffness from contact and non-contact contribution
K = Ka+Kns+Kts+Kac+Kbt+Kpst;

% --------------------------------------------------------------------------
% Evaluate tendon jacking reactions
% Calculate displacement increase vector d_F
if ODB(step).anchor == 0
   K_EF = K(nd+1:neq,1:nd);
   K_F = K(nd+1:neq,nd+1:neq);
   f_F = residual_F;
```

```
    d_F = K_F\(f_F-K_EF*d_E);

    % Determine if it is the first loop for displacement increase vector
    % d_E
    d_Ee = zeros(nd,1);

    % Calculate residual force vector f_F
    % Calculate current displacement U = U+d_F
    Ue = U+[d_Ee;d_F];

    % Update current node position
    for i = 1:nnd
        xe(i) = xi(i)+U(ID(ID2(i,1)));
        ye(i) = yi(i)+U(ID(ID2(i,2)));
    end

    [F_c,slip2] =
Assembly_F_ci(epsilon,xe,ye,xi,yi,ID,ID2,elec,neq,slip,Ue,Mu);

    Kps = Assembly_kps(xi,yi,At,Lt,elet,neq,ID,ID2,Eps,U,ODB);
    Kb =
Assembly_kb(ID,ID2,neq,xi,yi,eleb,Lb,Ue,concrete_stress,concrete_strain
,A,S,I,hf,n_fiber,...
    steel_stress,steel_strain,zs,As,Ss,Is,Ec,Es);
    % Update residual force vector from non-contact contribution
    F_nc = (Ka+Kb+Kps)*Ue;
    %F = -F_ac+F_c+F_nc;
    F = F_c+F_nc;

    % Update equavilent force from tendon jacking appled at beam
    F_jacking1 = zeros(neq,1);
    for i = 1:size(eleac,1)
        F_jacking1(ID(ID2(eleac(i,2),1))) = -F_nc(ID(ID2(eleac(i,1),1)));
        F_jacking1(ID(ID2(eleac(i,2),2))) = -F_nc(ID(ID2(eleac(i,1),2)));
    end
    residual_F = residual_Fi+F_jacking1(nd+1:neq);
else
    F_jacking1 = zeros(neq,1);
end

% Loop2-------------------------------------------------------------------
```

```
% Newton-Rapshon solver
normf = 1;
normfinc = 1;
normtest = 2;
normtest2 = 1;
flag2 = 0;
count2 = 1;
% eng_inter_I = 0;
acur_ctrl = 0.001;
F_norm_ratio = acur_ctrl+1;
Disp_norm_ratio = acur_ctrl+1;
% eng_ratio = acur_ctrl+0.1;
Mu = ODB(step).Material.Mu;

while (F_norm_ratio >= acur_ctrl || Disp_norm_ratio >= acur_ctrl) &&
count2 <= 150 && normf >= 1%&& normfinc > 0
    if count2 > 1
        d_E = zeros(nd,1);
    end

    % Calculate displacement increase vector d_F
    K_EF = K(nd+1:neq,1:nd);
    K_F = K(nd+1:neq,nd+1:neq);
    f_F = residual_F;
    d_F = K_F\(f_F-K_EF*d_E);
    U = U+[d_E;d_F];

    % Update current node position
    for i = 1:nnd
        x(i) = xi(i)+U(ID(ID2(i,1)));
        y(i) = yi(i)+U(ID(ID2(i,2)));
    end

    if ODB(step).anchor == 0
        Kac = zeros(neq);
        F_ac = zeros(neq,1);
    else
        % Update current global stiffness matrix Kac from anchorage
        Kac = Assembly_kac(eleac,epsilon2,ID,ID2,neq);
        % Update residual force vector from anchorage contribution
```

```
    F_ac =
Assembly_F_aci(epsilon2,x,y,anchor_u,anchor_v,ID,ID2,eleac,neq);
  end

  % Update current global stiffness matrix Kns & Kts from contact
contribution
  % tangent stiffness matrix from normal contact
  Kns = Assembly_kns(epsilon,x,y,xi,yi,ID,ID2,elec,neq);
  % tangetn stiffness matrix from tagent contact
  Kts = Assembly_kts2(epsilon,x,y,xi,yi,ID,ID2,elec,neq,Mu,slip);
  % Update residual force vector from contact contribution
  [F_c,slip2] =
Assembly_F_ci(epsilon,x,y,xi,yi,ID,ID2,elec,neq,slip,U,Mu);

  Kpst = Assembly_kpst(xi,yi,At,Lt,elet,neq,ID,ID2,Epst,U);
  Kps = Assembly_kps(xi,yi,At,Lt,elet,neq,ID,ID2,Eps,U,ODB);

  % Update tangential stiffness matrix from nonlinear beam element
  % Update stiffness matrix from nonlinear beam element
  Kbt =
Assembly_kbt(ID,ID2,neq,xi,yi,eleb,Lb,U,concrete_stress,concrete_strain,
A,S,I,hf,n_fiber,...
  steel_stress,steel_strain,zs,As,Ss,Is,Ect,Est);
  Kb =
Assembly_kb(ID,ID2,neq,xi,yi,eleb,Lb,U,concrete_stress,concrete_strain,
A,S,I,hf,n_fiber,...
  steel_stress,steel_strain,zs,As,Ss,Is,Ec,Es);

  % Update residual force vector from non-contact contribution
  F_nc = (Ka+Kb+Kps)*U;
  %F = -F_ac+F_c+F_nc;
  F = -F_ac+F_c+F_nc;

  % Update equavilent force from tendon jacking appled at beam
  F_jacking2 = zeros(neq,1);
  if ODB(step).anchor == 0
    for i = 1:size(eleac,1)
      F_jacking2(ID(ID2(eleac(i,2),1))) = -F_nc(ID(ID2(eleac(i,1),1)));
      F_jacking2(ID(ID2(eleac(i,2),2))) = -F_nc(ID(ID2(eleac(i,1),2)));
    end
  end
```

```
   residual_F = -F(nd+1:neq)+residual_Fi+F_jacking1(nd+1:neq);
   K = Ka+Kns+Kts+Kac+Kbt+Kpst;

   % Convergence criteria
   F_inter_norm = norm(F+F_ac);
   F_err_norm = norm(residual_F);
   Disp_all_norm = norm(U(nd+1:neq));
   Disp_inc_norm = norm(d_F);
   F_norm_ratio = F_err_norm/F_inter_norm;                    % monitor
this ratio while iteration
   if Disp_inc_norm == 0 || (Disp_inc_norm == Disp_all_norm &&
step >= 2)
      Disp_norm_ratio = 0.9*acur_ctrl;
      F_norm_ratio = 0.9*acur_ctrl;
   else
      Disp_norm_ratio = Disp_inc_norm/Disp_all_norm;           %
monitor this ratio while iteration
   end

   F_jacking3 = F_jacking2-F_jacking1;
   residual_Ftrial = residual_F-F_jacking3(nd+1:neq);
   normf = norm(residual_F)
   if count2 > 1
   normfinc = normf2 - normf;
   end
   normf2 = normf;
   normtest2 = normtest;
   if norm(residual_F)>0
      residual_F = residual_Ftrial-F_jacking3(nd+1:neq);
   end
   F_jacking1 = F_jacking2;
   count2 = count2+1
end;

if count2 < 150 && (normfinc > 0 || F_norm_ratio < acur_ctrl || normf <
100)%&& normtest/normf > .001 %%&& normtest < 5000
   % Transfer U residual_F from new node ordering to origional ordering
   Uo = Reorder2(ID,U);
   residual_F2 = -F+[zeros(nd,1);residual_Fi];
   residual_Fo = Reorder2(ID,residual_F2);
```

```
   % Update new residual froce and displacement vector to model database
(ODB)
   % Update initial displacement vector
   ODB(step+1) = ODB(step);
   ODB(step+1).slip = slip2;
   ODB = ODBUpdate(ODB,step,Uo,residual_Fo);
end
```

Solver for frictional contact (partially unbonded and bonded condition)

MainSolver2.m

```
% Solver main function for frictional contact
function ODB = MainSolver2(ODB,step)
xi = ODB(step).Node.X;
yi = ODB(step).Node.Y;
Et = ODB(step).TrussElement.E(1);
At = ODB(step).TrussElement.A(1);
Eemb = ODB(step).EmbeddingElement.E(1);
Aemb = ODB(step).EmbeddingElement.A(1);
Iemb = ODB(step).EmbeddingElement.I(1);
Lt = ODB(step).TrussElement.Length;
Lb = ODB(step).BeamElement.Length;
Lemb = ODB(step).EmbeddingElement.Length;
elet = ODB(step).TrussElement.Nodes;
eleb = ODB(step).BeamElement.Nodes;
eleac = ODB(step).AnchorElement.Nodes;
elec = ODB(step).ContactElement.Nodes;
eleemb = ODB(step).EmbeddingElement.Nodes;
neq = ODB(step).neq;
nd = ODB(step).nd;
nnd = ODB(step).nnd;
ID = ODB(step).ID;
d_E = ODB(step).d_E;
ID2 = ODB(step).ID2;
slip = ODB(step).slip;
slip2 = slip;
Mu = ODB(step).Material.Mu;
Uo = InitialCondition(ODB(step).Node.U,ODB(step).Node.V,...
   ODB(step).Node.T,ODB(step).Node.Boundary);
residual_Fo = InitialCondition(ODB(step).Node.Fx_residual,...
```

```
ODB(step).Node.Fy_residual,ODB(step).Node.M_residual,ODB(step).No
de.Boundary);
epsilon = ODB(step).ContactElement.Epsilon;
epsilon2 = ODB(step).AnchorElement.Epsilon;
x = xi;
y = yi;

% concrete unaxial stress-strain
concrete_stress = ODB(step).Material.Concrete(:,1)';
concrete_strain = ODB(step).Material.Concrete(:,2)';

% concrete beam section geometry
n_fiber = ODB(step).BeamElement.n;
hf = ODB(step).BeamElement.hf;
A = ODB(step).BeamElement.A;
S = ODB(step).BeamElement.S;
I = ODB(step).BeamElement.I;

% non-prestressed steel unaxial stress-strain
steel_stress = ODB(step).Material.Steel(:,1)';
steel_strain = ODB(step).Material.Steel(:,2)';

Ec = ODB(step).Material.Ec;
Es = ODB(step).Material.Es;
Eps = ODB(step).Material.Eps;
Ect = ODB(step).Material.Ect;
Est = ODB(step).Material.Est;
Epst = ODB(step).Material.Epst;

% non-prestressed steel geometry
zs = ODB(step).BeamElement.zs;
As = ODB(step).BeamElement.As;
Ss = ODB(step).BeamElement.Ss;
Is = ODB(step).BeamElement.Is;

% Determine if at anchorage stage and caluclate relative disp between two
% nodes before prestressing
if ODB(step).anchor == 1 && step == 2
   for i = 1:size(eleac,1)
      anchor_u(i) = x(eleac(i,2))+Uo(ID2(eleac(i,2),1))...
```

```
        -x(eleac(i,1))-Uo(ID2(eleac(i,1),1))
      anchor_v(i) = y(eleac(i,2))+Uo(ID2(eleac(i,2),2))...
        -y(eleac(i,1))-Uo(ID2(eleac(i,1),2))
   end
elseif ODB(step).anchor == 0
   anchor_u = zeros(size(eleac,1),1);
   anchor_v = zeros(size(eleac,1),1);
elseif step >= 3
   anchor_u = ODB(2).anchor_u;
   anchor_v = ODB(2).anchor_v;
end
ODB(step).anchor_u = anchor_u;
ODB(step).anchor_v = anchor_v;

% Transfer Uo residual_Fo from origional node ordering to new ordering
U = Reorder(ID,Uo);
residual_F = Reorder(ID,residual_Fo);
residual_F = residual_F(nd+1:neq);
residual_Fi = residual_F;

for i = 1:nnd
   x1(i) = xi(i)+U(ID(ID2(i,1)));
   y1(i) = yi(i)+U(ID(ID2(i,2)));
end

% tangent stiffness matrix from anchorage
if ODB(step).anchor == 0
   Kac = zeros(neq);
else
   Kac = Assembly_kac(eleac,epsilon,ID,ID2,neq);
end

% tangent stiffness matrix from normal contact
Kns = Assembly_kns(epsilon,x1,y1,xi,yi,ID,ID2,elec,neq);
% tangent stiffness matrix from tagent contact (slip)
Kts = Assembly_kts2(epsilon,x1,y1,xi,yi,ID,ID2,elec,neq,Mu,slip);

% Tangent stiffness matrix from linear elements (constant considering
elasticity and linear geometry)
```

```
Ka =
Assembly_ka(xi,yi,Et,At,Eemb,Aemb,Iemb,Lt,Lemb,elet,eleemb,neq,ID,I
D2,ODB);

% Tanget stiffness matrix from ptrestressed steel truss element
Kpst = Assembly_kpst(xi,yi,At,Lt,elet,neq,ID,ID2,Epst,U);

% Tangent stiffness matrix from nonlinear beam
Kbt =
Assembly_kbt(ID,ID2,neq,xi,yi,eleb,Lb,U,concrete_stress,concrete_strain,
A,S,I,hf,n_fiber,...
   steel_stress,steel_strain,zs,As,Ss,Is,Ect,Est);

% Sum up tagent stiffness from contact and non-contact contribution
K = Ka+Kns+Kts+Kac+Kbt+Kpst;

% Evaluate tendon jacking reactions
% Calculate displacement increase vector d_F
if ODB(step).anchor == 0
   K_EF = K(nd+1:neq,1:nd);
   K_F = K(nd+1:neq,nd+1:neq);
   f_F = residual_F;
   d_F = K_F\(f_F-K_EF*d_E);

   % Determine if it is the first loop for displacement increase vector
   d_Ee = zeros(nd,1);

   % Calculate residual force vector f_F
   % Calculate current displacement U = U+d_F
   Ue = U+[d_Ee;d_F];

   % Update current node position
   for i = 1:nnd
      xe(i) = xi(i)+U(ID(ID2(i,1)));
      ye(i) = yi(i)+U(ID(ID2(i,2)));
   end

   [F_c,slip2] =
Assembly_F_ci(epsilon,xe,ye,xi,yi,ID,ID2,elec,neq,slip,Ue,Mu);
   Kps = Assembly_kps(xi,yi,At,Lt,elet,neq,ID,ID2,Eps,U,ODB);
```

```
    Kb =
Assembly_kb(ID,ID2,neq,xi,yi,eleb,Lb,Ue,concrete_stress,concrete_strain
,A,S,I,hf,n_fiber,...
    steel_stress,steel_strain,zs,As,Ss,Is,Ec,Es);
    % Update residual force vector from non-contact contribution
    F_nc = (Ka+Kb+Kps)*Ue;
    F = F_c+F_nc;

    % Update equavilent force from tendon jacking appled at beam
    F_jacking1 = zeros(neq,1);
    for i = 1:size(eleac,1)
       F_jacking1(ID(ID2(eleac(i,2),1))) = -F_nc(ID(ID2(eleac(i,1),1)));
       F_jacking1(ID(ID2(eleac(i,2),2))) = -F_nc(ID(ID2(eleac(i,1),2)));
    end
    residual_F = residual_Fi+F_jacking1(nd+1:neq);
else
    F_jacking1 = zeros(neq,1);
end

% tangent stiffness matrix from tagent contact (slip)
Kts = Assembly_kts3(epsilon,x1,y1,xi,yi,ID,ID2,elec,neq,Mu,slip);

% Sum up tagent stiffness from contact and non-contact contribution
K = Ka+Kns+Kts+Kac+Kbt+Kpst;

% Newton-Rapshon solver
iter_loop1 = 50;
iter_loop2 = 20;
residual_Ftemp = residual_F;
count5 = 1;
acur_ctrl_min = 0.001;
normf = 1;
acur_ctrl = acur_ctrl_min;
F_norm_ratio = acur_ctrl+1;
Disp_norm_ratio = acur_ctrl+1;

while (F_norm_ratio >= acur_ctrl || Disp_norm_ratio >= acur_ctrl) &&
count5 <= 10 && normf >= 1 %8.9.2016 count5 <= 10 chagne to <=1 to
disable kts4 loop
    normf = 1;
    normfinc = 1;
```

```
normtest = 2;
normtest2 = 1;
flag2 = 0;
count2 = 1;
acur_ctrl = acur_ctrl_min+0.001*(count5-1);
F_norm_ratio = acur_ctrl+1;
Disp_norm_ratio = acur_ctrl+1;

if count5 < 3 %count5 loop used to loose convergence criteria
   while (F_norm_ratio >= acur_ctrl || Disp_norm_ratio >= acur_ctrl)
&& count2 <= iter_loop1 && normf >= 1%&& normfinc > 0
      if count2 > 1
         d_E = zeros(nd,1);
      end
      K_EF = K(nd+1:neq,1:nd);
      K_F = K(nd+1:neq,nd+1:neq);
      f_F = residual_F;
      d_F = K_F\(f_F-K_EF*d_E);
      U = U+[d_E;d_F];
      % Update current node position
      for i = 1:nnd
         x(i) = xi(i)+U(ID(ID2(i,1)));
         y(i) = yi(i)+U(ID(ID2(i,2)));
      end

      if ODB(step).anchor == 0
         Kac = zeros(neq);
         F_ac = zeros(neq,1);
      else
         % Update current global stiffness matrix Kac from anchorage
         Kac = Assembly_kac(eleac,epsilon2,ID,ID2,neq);
         % Update residual force vector from anchorage contribution
         F_ac =
Assembly_F_aci(epsilon2,x,y,anchor_u,anchor_v,ID,ID2,eleac,neq);
      end

      % Update current global stiffness matrix Kns & Kts from contact
contribution
      % tangent stiffness matrix from normal contact
      Kns = Assembly_kns(epsilon,x,y,xi,yi,ID,ID2,elec,neq);
      % Update residual force vector from contact contribution
```

```
        [F_c,slip2] =
Assembly_F_ci2(epsilon,x,y,xi,yi,ID,ID2,elec,neq,slip,slip2,U,Mu);

        % tangetn stiffness matrix from tagent contact

        Kts =
Assembly_kts2(epsilon,x,y,xi,yi,ID,ID2,elec,neq,Mu,slip,slip2,count2);

        Kpst = Assembly_kpst(xi,yi,At,Lt,elet,neq,ID,ID2,Epst,U);
        Kps = Assembly_kps(xi,yi,At,Lt,elet,neq,ID,ID2,Eps,U,ODB);

        % Update tangential stiffness matrix from nonlinear beam element
        % Update stiffness matrix from nonlinear beam element
        Kbt =
Assembly_kbt(ID,ID2,neq,xi,yi,eleb,Lb,U,concrete_stress,concrete_strain,
A,S,I,hf,n_fiber,...
        steel_stress,steel_strain,zs,As,Ss,Is,Ect,Est);
        Kb =
Assembly_kb(ID,ID2,neq,xi,yi,eleb,Lb,U,concrete_stress,concrete_strain,
A,S,I,hf,n_fiber,...
        steel_stress,steel_strain,zs,As,Ss,Is,Ec,Es);

        F_nc = (Ka+Kb+Kps)*U;
        F = -F_ac+F_c+F_nc;

        % Update equavilent force from tendon jacking applied at beam
        F_jacking2 = zeros(neq,1);
        if ODB(step).anchor == 0
           for i = 1:size(eleac,1)
              F_jacking2(ID(ID2(eleac(i,2),1))) = -
F_nc(ID(ID2(eleac(i,1),1)));
              F_jacking2(ID(ID2(eleac(i,2),2))) = -
F_nc(ID(ID2(eleac(i,1),2)));
           end
        end

        residual_F = -F(nd+1:neq)+residual_Fi+F_jacking1(nd+1:neq);
        K = Ka+Kns+Kts+Kac+Kbt+Kpst;

        % Convergence creteria
        F_inter_norm = norm(F+F_ac);
```

```
        F_err_norm = norm(residual_F);
        Disp_all_norm = norm(U(nd+1:neq));
        Disp_inc_norm = norm(d_F);
        F_norm_ratio = F_err_norm/F_inter_norm; % monitor this ratio
while iteration
        if Disp_inc_norm == 0 || (Disp_inc_norm == Disp_all_norm &&
step >= 2)
            Disp_norm_ratio = 0.9*acur_ctrl;
            F_norm_ratio = 0.9*acur_ctrl;
        else
            Disp_norm_ratio = Disp_inc_norm/Disp_all_norm; % monitor
this ratio while iteration
        end

        % search optimal residual force factor
        F_jacking3 = F_jacking2-F_jacking1;

        if count2 <= iter_loop1
            count4 = 0;
            normtest = norm(residual_F)+1;
            while normtest >= norm(residual_F) && count4 < 4 &&
norm(residual_F)>0 || isfinite(normtest) == 0
                residual_Ftrial = 1/(10^count4)*residual_F-
F_jacking3(nd+1:neq);
                if normtest2 < 1 || flag2 == 1;
                    residual_Ftest =
Residual_Ftest(residual_Ftrial,K,nd,neq,d_E,xi,yi,ODB,...

step,epsilon,ID,ID2,elec,slip,slip2,Mu,Ka,F_jacking2,U,nnd,residual_Fi,e
psilon2,anchor_u,anchor_v,eleac);
                flag2 = 1;
                else
                    residual_Ftest =
Residual_Ftest(residual_Ftrial,K,nd,neq,d_E,xi,yi,ODB,...

step,epsilon,ID,ID2,elec,slip,slip2,Mu,Ka,F_jacking2,U,nnd,residual_Fi,e
psilon2,anchor_u,anchor_v,eleac);
                end
                count4 = count4+1;
                normtest = norm(residual_Ftest);
            end
```

```
        else

            residual_Ftrial2 = residual_F;
            residual_Ftest = Residual_Ftest(residual_F-
F_jacking3(nd+1:neq),K,nd,neq,d_E,xi,yi,ODB,...

step,epsilon,ID,ID2,elec,slip,slip2,Mu,Ka,F_jacking2,U,nnd,residual_Fi,e
psilon2,anchor_u,anchor_v,eleac);
            flag = 0;
            count3 = 1;
            residual_l = 0;
            residual_r = residual_Ftrial2;
            while flag < 1
                residual_Ftest_r = Residual_Ftest(1.1*residual_Ftrial2-
F_jacking3(nd+1:neq),K,nd,neq,d_E,xi,yi,ODB,...

step,epsilon,ID,ID2,elec,slip,slip2,Mu,Ka,F_jacking2,U,nnd,residual_Fi,e
psilon2,anchor_u,anchor_v,eleac);
                residual_Ftest_l = Residual_Ftest(0.9*residual_Ftrial2-
F_jacking3(nd+1:neq),K,nd,neq,d_E,xi,yi,ODB,...

step,epsilon,ID,ID2,elec,slip,slip2,Mu,Ka,F_jacking2,U,nnd,residual_Fi,e
psilon2,anchor_u,anchor_v,eleac);
                normtest_r = norm(residual_Ftest_r);
                normtest_l = norm(residual_Ftest_l);
                normtest = norm(residual_Ftest);
                if normtest_r > normtest && normtest > normtest_l
                    residual_r = residual_Ftrial2;
                    residual_Ftest =
Residual_Ftest(0.5*(residual_l+residual_r)-
F_jacking3(nd+1:neq),K,nd,neq,d_E,xi,yi,ODB,...

step,epsilon,ID,ID2,elec,slip,slip2,Mu,Ka,F_jacking2,U,nnd,residual_Fi,e
psilon2,anchor_u,anchor_v,eleac);
                    residual_Ftrial = 0.5*(residual_l+residual_r)-
F_jacking3(nd+1:neq);
                    residual_Ftrial2 = residual_Ftrial+F_jacking3(nd+1:neq);
                elseif normtest_r < normtest && normtest < normtest_l
                    if count3 == 1
                        count1 = 1;
                        while normtest_r < normtest && normtest < normtest_l
```

```
                    residual_r = 10^count1*residual_F;
                    residual_Ftest = Residual_Ftest(residual_r-
F_jacking3(nd+1:neq),K,nd,neq,d_E,xi,yi,ODB,...

step,epsilon,ID,ID2,elec,slip,slip2,Mu,Ka,F_jacking2,U,nnd,residual_Fi,e
psilon2,anchor_u,anchor_v,eleac);
                    residual_Ftest_r = Residual_Ftest(1.1*residual_r-
F_jacking3(nd+1:neq),K,nd,neq,d_E,xi,yi,ODB,...

step,epsilon,ID,ID2,elec,slip,slip2,Mu,Ka,F_jacking2,U,nnd,residual_Fi,e
psilon2,anchor_u,anchor_v,eleac);
                    residual_Ftest_l = Residual_Ftest(0.9*residual_r-
F_jacking3(nd+1:neq),K,nd,neq,d_E,xi,yi,ODB,...

step,epsilon,ID,ID2,elec,slip,slip2,Mu,Ka,F_jacking2,U,nnd,residual_Fi,e
psilon2,anchor_u,anchor_v,eleac);
                    normtest_r = norm(residual_Ftest_r);
                    normtest_l = norm(residual_Ftest_l);
                    normtest = norm(residual_Ftest);
                    count1 = count1+1;
                end
                residual_Ftest =
Residual_Ftest(0.5*(residual_F+residual_r)-
F_jacking3(nd+1:neq),K,nd,neq,d_E,xi,yi,ODB,...

step,epsilon,ID,ID2,elec,slip,slip2,Mu,Ka,F_jacking2,U,nnd,residual_Fi,e
psilon2,anchor_u,anchor_v,eleac);
                residual_Ftrial = 0.5*(residual_F+residual_r)-
F_jacking3(nd+1:neq);
                residual_Ftrial2 = residual_Ftrial+F_jacking3(nd+1:neq);
            else
                residual_l = residual_Ftrial2;
                residual_Ftest =
Residual_Ftest(0.5*(residual_l+residual_r)-
F_jacking3(nd+1:neq),K,nd,neq,d_E,xi,yi,ODB,...

step,epsilon,ID,ID2,elec,slip,slip2,Mu,Ka,F_jacking2,U,nnd,residual_Fi,e
psilon2,anchor_u,anchor_v,eleac);
                residual_Ftrial = 0.5*(residual_l+residual_r)-
F_jacking3(nd+1:neq);
                residual_Ftrial2 = residual_Ftrial+F_jacking3(nd+1:neq);
```

```
            end
          else
            flag = 1;
          end
          count3 = count3+1;
        end
      end

      normtest = norm(residual_Ftest);
      normf = norm(residual_F);

      if count2 > 1
      normfinc = normf2 - normf;
      end
      normf2 = normf
      normtest2 = normtest;

      if norm(residual_F)>0
        residual_F = residual_Ftrial+F_jacking3(nd+1:neq);
      end
      F_jacking1 = F_jacking2;
      count2 = count2+1
    end;

  elseif count5 >= 3
    while (F_norm_ratio >= acur_ctrl || Disp_norm_ratio >= acur_ctrl)
&& count2 <= iter_loop2 && normf >= 1
      if count2 > 1
        d_E = zeros(nd,1);
      end

      % Calculate displacement increase vector d_F
      K_EF = K(nd+1:neq,1:nd);
      K_F = K(nd+1:neq,nd+1:neq);
      f_F = residual_F;
      d_F = K_F\(f_F-K_EF*d_E);

      % Calculate residual force vector f_F
      % Calculate current displacement U = U+d_F
      U = U+[d_E;d_F];
       % Update current node position
```

```
for i = 1:nnd
    x(i) = xi(i)+U(ID(ID2(i,1)));
    y(i) = yi(i)+U(ID(ID2(i,2)));
end

if ODB(step).anchor == 0
    Kac = zeros(neq);
    F_ac = zeros(neq,1);
else
    % Update current global stiffness matrix Kac from anchorage
    Kac = Assembly_kac(eleac,epsilon2,ID,ID2,neq);
    % Update residual force vector from anchorage contribution
    F_ac =
Assembly_F_aci(epsilon2,x,y,anchor_u,anchor_v,ID,ID2,eleac,neq);
end

% Update current global stiffness matrix Kns & Kts from contact
contribution
% tangent stiffness matrix from normal contact
Kns = Assembly_kns(epsilon,x,y,xi,yi,ID,ID2,elec,neq);

% Update residual force vector from contact contribution
[F_c,slip2] =
Assembly_F_ci3(epsilon,x,y,xi,yi,ID,ID2,elec,neq,slip,slip2,U,Mu);

% tangetn stiffness matrix from tagent contact
Kts =
Assembly_kts2(epsilon,x,y,xi,yi,ID,ID2,elec,neq,Mu,slip,slip2,count2);

% Tangent stiffness matrix from truss (constant considering
elasticity and linear geometry)
Kpst = Assembly_kpst(xi,yi,At,Lt,elet,neq,ID,ID2,Epst,U);
Kps = Assembly_kps(xi,yi,At,Lt,elet,neq,ID,ID2,Eps,U,ODB);

% Update tangential stiffness matrix from nonlinear beam element
% Update stiffness matrix from nonlinear beam element
Kbt =
Assembly_kbt(ID,ID2,neq,xi,yi,eleb,Lb,U,concrete_stress,concrete_strain,
A,S,I,hf,n_fiber,...
steel_stress,steel_strain,zs,As,Ss,Is,Ect,Est);
```

```
        Kb =
Assembly_kb(ID,ID2,neq,xi,yi,eleb,Lb,U,concrete_stress,concrete_strain,
A,S,I,hf,n_fiber,...
        steel_stress,steel_strain,zs,As,Ss,Is,Ec,Es);

        % Update residual force vector from non-contact contribution
        F_nc = (Ka+Kb+Kps)*U;
        F = -F_ac+F_c+F_nc;

        % Update equavilent force from tendon jacking appled at beam
        F_jacking2 = zeros(neq,1);
        if ODB(step).anchor == 0
          for i = 1:size(eleac,1)
            F_jacking2(ID(ID2(eleac(i,2),1))) = -
F_nc(ID(ID2(eleac(i,1),1)));
            F_jacking2(ID(ID2(eleac(i,2),2))) = -
F_nc(ID(ID2(eleac(i,1),2)));
          end
        end
        residual_F = -F(nd+1:neq)+residual_Fi+F_jacking1(nd+1:neq);
        K = Ka+Kns+Kts+Kac+Kbt+Kpst;

        % Convergence creteria
        F_inter_norm = norm(F+F_ac);
        F_err_norm = norm(residual_F);
        Disp_all_norm = norm(U(nd+1:neq));
        Disp_inc_norm = norm(d_F);
        F_norm_ratio = F_err_norm/F_inter_norm; % monitor this ratio
while iteration
        if Disp_inc_norm == 0 || (Disp_inc_norm == Disp_all_norm &&
step >= 2)
          Disp_norm_ratio = 0.9*acur_ctrl;
          F_norm_ratio = 0.9*acur_ctrl;
        else
          Disp_norm_ratio = Disp_inc_norm/Disp_all_norm; % monitor
this ratio while iteration
        end

        % search optimal residual force factor
        F_jacking3 = F_jacking2-F_jacking1;
        if count2 <= iter_loop2
```

```
        count4 = 0;
        normtest = norm(residual_F)+1;
        while normtest >= norm(residual_F) && count4 < 4 &&
norm(residual_F)>0 || isfinite(normtest) == 0
          residual_Ftrial = 1/(10^count4)*residual_F-
F_jacking3(nd+1:neq);
          if normtest2 < 1 || flag2 == 1;
            residual_Ftest =
Residual_Ftest(residual_Ftrial,K,nd,neq,d_E,xi,yi,ODB,...

step,epsilon,ID,ID2,elec,slip,slip2,Mu,Ka,F_jacking2,U,nnd,residual_Fi,e
psilon2,anchor_u,anchor_v,eleac);
          flag2 = 1;
          else
            residual_Ftest =
Residual_Ftest(residual_Ftrial,K,nd,neq,d_E,xi,yi,ODB,...

step,epsilon,ID,ID2,elec,slip,slip2,Mu,Ka,F_jacking2,U,nnd,residual_Fi,e
psilon2,anchor_u,anchor_v,eleac);
          end
          count4 = count4+1;
          normtest = norm(residual_Ftest);
        end
      else

      % line search
        residual_Ftrial2 = residual_F;
        residual_Ftest = Residual_Ftest(residual_F-
F_jacking3(nd+1:neq),K,nd,neq,d_E,xi,yi,ODB,...

step,epsilon,ID,ID2,elec,slip,slip2,Mu,Ka,F_jacking2,U,nnd,residual_Fi,e
psilon2,anchor_u,anchor_v,eleac);
        flag = 0;
        count3 = 1;
        residual_l = 0;
        residual_r = residual_Ftrial2;
        while flag < 1
          residual_Ftest_r = Residual_Ftest(1.1*residual_Ftrial2-
F_jacking3(nd+1:neq),K,nd,neq,d_E,xi,yi,ODB,...
```

```
step,epsilon,ID,ID2,elec,slip,slip2,Mu,Ka,F_jacking2,U,nnd,residual_Fi,e
psilon2,anchor_u,anchor_v,eleac);
          residual_Ftest_l = Residual_Ftest(0.9*residual_Ftrial2-
F_jacking3(nd+1:neq),K,nd,neq,d_E,xi,yi,ODB,...

step,epsilon,ID,ID2,elec,slip,slip2,Mu,Ka,F_jacking2,U,nnd,residual_Fi,e
psilon2,anchor_u,anchor_v,eleac);
          normtest_r = norm(residual_Ftest_r);
          normtest_l = norm(residual_Ftest_l);
          normtest = norm(residual_Ftest);
          if normtest_r > normtest && normtest > normtest_l
            residual_r = residual_Ftrial2;
            residual_Ftest =
Residual_Ftest(0.5*(residual_l+residual_r)-
F_jacking3(nd+1:neq),K,nd,neq,d_E,xi,yi,ODB,...

step,epsilon,ID,ID2,elec,slip,slip2,Mu,Ka,F_jacking2,U,nnd,residual_Fi,e
psilon2,anchor_u,anchor_v,eleac);
            residual_Ftrial = 0.5*(residual_l+residual_r)-
F_jacking3(nd+1:neq);
            residual_Ftrial2 = residual_Ftrial+F_jacking3(nd+1:neq);
          elseif normtest_r < normtest && normtest < normtest_l
            if count3 == 1
              count1 = 1;
              while normtest_r < normtest && normtest < normtest_l
                residual_r = 10^count1*residual_F;
                residual_Ftest = Residual_Ftest(residual_r-
F_jacking3(nd+1:neq),K,nd,neq,d_E,xi,yi,ODB,...

step,epsilon,ID,ID2,elec,slip,slip2,Mu,Ka,F_jacking2,U,nnd,residual_Fi,e
psilon2,anchor_u,anchor_v,eleac);
                residual_Ftest_r = Residual_Ftest(1.1*residual_r-
F_jacking3(nd+1:neq),K,nd,neq,d_E,xi,yi,ODB,...

step,epsilon,ID,ID2,elec,slip,slip2,Mu,Ka,F_jacking2,U,nnd,residual_Fi,e
psilon2,anchor_u,anchor_v,eleac);
                residual_Ftest_l = Residual_Ftest(0.9*residual_r-
F_jacking3(nd+1:neq),K,nd,neq,d_E,xi,yi,ODB,...

step,epsilon,ID,ID2,elec,slip,slip2,Mu,Ka,F_jacking2,U,nnd,residual_Fi,e
psilon2,anchor_u,anchor_v,eleac);
```

```
                normtest_r = norm(residual_Ftest_r);
                normtest_l = norm(residual_Ftest_l);
                normtest = norm(residual_Ftest);
                count1 = count1+1;
              end
              residual_Ftest =
Residual_Ftest(0.5*(residual_F+residual_r)-
F_jacking3(nd+1:neq),K,nd,neq,d_E,xi,yi,ODB,...

step,epsilon,ID,ID2,elec,slip,slip2,Mu,Ka,F_jacking2,U,nnd,residual_Fi,e
psilon2,anchor_u,anchor_v,eleac);
              residual_Ftrial = 0.5*(residual_F+residual_r)-
F_jacking3(nd+1:neq);
              residual_Ftrial2 = residual_Ftrial+F_jacking3(nd+1:neq);
            else
              residual_l = residual_Ftrial2;
              residual_Ftest =
Residual_Ftest(0.5*(residual_l+residual_r)-
F_jacking3(nd+1:neq),K,nd,neq,d_E,xi,yi,ODB,...

step,epsilon,ID,ID2,elec,slip,slip2,Mu,Ka,F_jacking2,U,nnd,residual_Fi,e
psilon2,anchor_u,anchor_v,eleac);
              residual_Ftrial = 0.5*(residual_l+residual_r)-
F_jacking3(nd+1:neq);
              residual_Ftrial2 = residual_Ftrial+F_jacking3(nd+1:neq);
            end
          else
            flag = 1;
          end
          count3 = count3+1;
        end
      end

      normtest = norm(residual_Ftest)
      normf = norm(residual_F)

      if count2 > 1
      normfinc = normf2 - normf;
      end
      normf2 = normf;
      normtest2 = normtest;
```

```
            if norm(residual_F)>0
                residual_F = residual_Ftrial+F_jacking3(nd+1:neq);
            end
            F_jacking1 = F_jacking2;
            count2 = count2+1
        end
    end
    count5 = count5+1
end


if count2 < iter_loop2+100000 && (normfinc > 0 || F_norm_ratio <
acur_ctrl || normf < 25)
    % Transfer U residual_F from new node ordering to origional ordering
    Uo = Reorder2(ID,U);
    residual_F2 = -F+[zeros(nd,1);residual_Fi];
    residual_Fo = Reorder2(ID,residual_F2);
    ODB(step+1) = ODB(step);
    ODB(step+1).slip = slip2;
    ODB = ODBUpdate(ODB,step,Uo,residual_Fo);
end
```

Na_matrixBeam.m

```
% Shape functions of axial deformation in the natural coordinate s
 function Na = Na_matrixBeam(s)
     Na(1)=0.5*(1-s);
     Na(2)=0;
     Na(3)=0;
     Na(4)=0.5*(s+1);
     Na(5)=0;
     Na(6)=0;
```

NmatrixBeam.m

```
% Shape functions in the natural coordinate s
 function N = NmatrixBeam(s,L)
     N(1)=0;
     N(2)=1/4*(1-s)^2*(2+s);
     N(3)=L/8*(1-s)^2*(1+s);
     N(4)=0;
```

```
    N(5)=1/4*(1+s)^2*(2-s);
    N(6)=L/8*(1+s)^2*(s-1);
```

NmatrixBeam1.m

```
% Shape functions in the natural coordinate s
 function N = NmatrixBeam1(s,L)
    N(1)=0;
    N(2)=1/4*(3*s^2-3);
    N(3)=L/8*(3*s^2-2*s+1);
    N(4)=0;
    N(5)=1/4*(-3*s^2+3);
    N(6)=L/8*(3*s^2+2*s-1);
```

NodeMapping.m

```
% Genereate NodeMapping array
function [ID,d] = NodeMapping(nnd,nd,dof,flags,e_bc)
count = 0;
count1 = 0;
count2 = 1;
for i = 1:nnd
   for j = 1:dof(i)
      if flags(i,j) == 2                    % check if essential boundary
        count = count + 1;
        ID(count2) = count;                 % number first the degrees-of-freedom
on essential boundary
        d(count)= e_bc(j,i);                % store the reordered values of
essential B.C
      elseif flags(i,j) == 1
        count1 = count1 + 1;
        ID(count2) = nd + count1;
      end
      count2 = count2 + 1;
   end
end
d = d';
```

NodeMapping2.m

```
% Genereate ID2 array
```

```
function ID2 = NodeMapping2(nnd,dof)
ID2 = zeros(nnd,3);
count = 0;
for i = 1:nnd
   if dof(i) == 2                    % check if node on slide line
      for j = 1:2
         ID2(i,j) = count+1;
         count = count+1;
      end
   else
      for j = 1:3
         ID2(i,j) = count+1;
         count = count+1;
      end
   end
end
```

ODBUpdate.m

```
% Update output database
function ODB = ODBUpdate(ODB,step,Uo,residual_Fo)
for i = 1 : ODB(step).nnd
   if ODB(step).ID2(i,1)>0 && ODB(step).ID2(i,2)>0 &&
ODB(step).ID2(i,3)>0
      ODB(step+1).Node.U(i) = Uo(ODB(step).ID2(i,1));
      ODB(step+1).Node.V(i) = Uo(ODB(step).ID2(i,2));
      ODB(step+1).Node.T(i) = Uo(ODB(step).ID2(i,3));
   else
      ODB(step+1).Node.U(i) = Uo(ODB(step).ID2(i,1));
      ODB(step+1).Node.V(i) = Uo(ODB(step).ID2(i,2));
      ODB(step+1).Node.T(i) = 0;
   end
end

for i = 1 : ODB(step).nnd
   if ODB(step).ID2(i,1)>0 && ODB(step).ID2(i,2)>0 &&
ODB(step).ID2(i,3)>0
      ODB(step+1).Node.Fx_residual(i) =
residual_Fo(ODB(step).ID2(i,1));
      ODB(step+1).Node.Fy_residual(i) =
residual_Fo(ODB(step).ID2(i,2));
```

```
      ODB(step+1).Node.M_residual(i) =
residual_Fo(ODB(step).ID2(i,3));
   else
      ODB(step+1).Node.Fx_residual(i) =
residual_Fo(ODB(step).ID2(i,1));
      ODB(step+1).Node.Fy_residual(i) =
residual_Fo(ODB(step).ID2(i,2));
      ODB(step+1).Node.M_residual(i) = 0;
   end
end
```

Output_fps.m

```
% Calculate stress in tendons.X
function fps = Output_fps(ODB,step)
L_old =
ElementLength(ODB(step).TrussElement.Nodes,ODB(step).Node.X,ODB
(step).Node.Y);
L_new =
ElementLength(ODB(step).TrussElement.Nodes,ODB(step).Node.X+OD
B(step).Node.U,ODB(step).Node.Y+ODB(step).Node.V);
for i = 1:size(ODB(step).TrussElement.Nodes,1)
   fps(i) = (L_new(i)-L_old(i))/L_old(i)*ODB(step).TrussElement.E(i);
end
```

Reorder.m

```
% Switch node ordering from initial oder to equation order
function newordering = Reorder(ID,oldordering)
for i = 1:size(oldordering,1)
   newordering(ID(i),1) = oldordering(i);
end
```

Reorder2.m

```
% Switch node ordering from equation order to initial order
function newordering = Reorder2(ID,oldordering)
for i = 1:size(oldordering,1)
   newordering(i,1) = oldordering(ID(i));
end
```

Residual_Ftest.m

```
% Nonlinear beam stiffness matrix is removed for speeding up
function residual_Ftest =
Residual_Ftest(residual_Ftrial,K,nd,neq,d_E,xi,yi,ODB,...

step,epsilon,ID,ID2,elec,slip,slip2,Mu,Ka,F_jacking2,U,nnd,residual_Fi,e
psilon2,anchor_u,anchor_v,eleac)
    % concrete unaxial stress-strain
    concrete_stress = ODB(step).Material.Concrete(:,1)';
    concrete_strain = ODB(step).Material.Concrete(:,2)';
    % extend stress-strain curve
    concrete_stress =
[concrete_stress(1),concrete_stress,concrete_stress(size(concrete_stress,2)
)];
    concrete_strain =
[10*concrete_strain(1),concrete_strain,10*concrete_strain(size(concrete_s
train,2))];

    % concrete beam section geometry
    n_fiber = ODB(step).BeamElement.n;
    hf = ODB(step).BeamElement.hf;
    A = ODB(step).BeamElement.A;
    S = ODB(step).BeamElement.S;
    I = ODB(step).BeamElement.I;

    % non-prestressed steel unaxial stress-strain
    steel_stress = ODB(step).Material.Steel(:,1)';
    steel_strain = ODB(step).Material.Steel(:,2)';
    % extend stress-strain curve
    steel_stress =
[steel_stress(1),steel_stress,steel_stress(size(steel_stress,2))];
    steel_strain =
[10*steel_strain(1),steel_strain,10*steel_strain(size(steel_strain,2))];

    % non-prestressed steel geometry
    zs = ODB(step).BeamElement.zs;
    As = ODB(step).BeamElement.As;
    Ss = ODB(step).BeamElement.Ss;
    Is = ODB(step).BeamElement.Is;

    Ec = ODB(step).Material.Ec;
```

```
Es = ODB(step).Material.Es;
Eps = ODB(step).Material.Eps;

Lb = ODB(step).BeamElement.Length;
eleb = ODB(step).BeamElement.Nodes;

Et = ODB(step).TrussElement.E(1);
At = ODB(step).TrussElement.A(1);
Eemb = ODB(step).EmbeddingElement.E(1);
Aemb = ODB(step).EmbeddingElement.A(1);
Iemb = ODB(step).EmbeddingElement.I(1);
Lt = ODB(step).TrussElement.Length;
Lemb = ODB(step).EmbeddingElement.Length;
elet = ODB(step).TrussElement.Nodes;
eleemb = ODB(step).EmbeddingElement.Nodes;

% Determine if it is the first loop for displacement increase vector
d_E = zeros(nd,1);

% Calculate displacement increase vector d_F
K_EF = K(nd+1:neq,1:nd);
K_F = K(nd+1:neq,nd+1:neq);
f_F = residual_Ftrial;
d_F = K_F\(f_F-K_EF*d_E);
d_Fftest = d_F;

% Calculate residual force vector f_F
% Calculate current displacement U = U+d_F
U2 = U+[d_E;d_F];

% Update current node position
for i = 1:nnd
   x2(i) = xi(i)+U2(ID(ID2(i,1)));
   y2(i) = yi(i)+U2(ID(ID2(i,2)));
end

% Update residual force vector from anchorage contribution
if ODB(step).anchor == 0
   F_ac = zeros(neq,1);
else
   % Update residual force vector from anchorage contribution
```

```
        F_ac =
Assembly_F_aci(epsilon2,x2,y2,anchor_u,anchor_v,ID,ID2,eleac,neq);
    end
    %F_ac =
Assembly_F_aci(epsilon2,x2,y2,anchor_u,anchor_v,ID,ID2,eleac,neq);
    [F_c,slip3] =
Assembly_F_ci2(epsilon,x2,y2,xi,yi,ID,ID2,elec,neq,slip,slip2,U2,Mu);

    % Tangent stiffness matrix from truss (constant considering elasticity
and linear geometry)
    Kps = Assembly_kps(x2,y2,At,Lt,elet,neq,ID,ID2,Eps,U,ODB);

    % Update stiffness maxtrix from nonlienar beam
    Kb =
Assembly_kb(ID,ID2,neq,x2,y2,eleb,Lb,U2,concrete_stress,concrete_strai
n,A,S,I,hf,n_fiber,...
        steel_stress,steel_strain,zs,As,Ss,Is,Ec,Es);
    % Update residual force vector from non-contact contribution
    F_nc = (Ka+Kb+Kps)*U2;
    F = -F_ac+F_c+F_nc;

    % Update equavilent force from tendon jacking appled at beam
    F_jacking2 = zeros(neq,1);
    if ODB(step).anchor == 0
        for i = 1:size(eleac,1)
            F_jacking2(ID(ID2(eleac(i,2),1))) = -
F_nc(ID(ID2(eleac(i,1),1)));
            F_jacking2(ID(ID2(eleac(i,2),2))) = -
F_nc(ID(ID2(eleac(i,1),2)));
        end
    end

    residual_Ftest = -F(nd+1:neq)+residual_Fi+F_jacking2(nd+1:neq);
```

Slip_array.m

```
% Initiate node-slip array
function slip = Slip_array(ODB,step)

for i = 1:size(ODB(step).ContactElement.Nodes,1)
    slip(i,1) = ODB(step).ContactElement.Nodes(i,1);
```

```
  slip(i,2) = 0;
end
```

slopinterpolator.m

```
%stress-strain intepoator
function [E,Et] = slopinterpolator(stress,strain,stress0,strain0,E0)
E = zeros(2,size(stress,2));
Et = zeros(2,size(stress,2));

for loop1 = 1:2
  for loop2 = 1:size(stress,2)
    if strain(loop1,loop2) == 0
      E(loop1,loop2) = E0;
      Et(loop1,loop2) = E0;
    else
      E(loop1,loop2) = stress(loop1,loop2)/strain(loop1,loop2);
      Et(loop1,loop2) =
(interp1(strain0,stress0,strain(loop1,loop2)+0.000001)...
          -interp1(strain0,stress0,strain(loop1,loop2)-
0.000001))/0.000002;
    end
  end
end
```

SmatrixBeam.m

```
% Second derivative of the shape functions with repsect to 's' in the natural
coordinate s
function B = SmatrixBeam(s,L)
   B(1)=0;
   B(2)=3/2*s;
   B(3)=L*(3/4*s-1/4);
   B(4)=0;
   B(5)=-3/2*s;
   B(6)=L*(3/4*s+1/4);
```

Transform.m

```
% Coordinates transformation matrix
function T = TransformB(a)
```

```
    T = [cos(a),sin(a),0,0,0,0;-
sin(a),cos(a),0,0,0,0;0,0,1,0,0,0;0,0,0,cos(a),sin(a),0;0,0,0,-
sin(a),cos(a),0;0,0,0,0,0,1];
```

TransformB.m

```
% Coordinates transformation matrix
function T = TransformB(a)
    T = [cos(a),sin(a),0,0,0,0;-
sin(a),cos(a),0,0,0,0;0,0,1,0,0,0;0,0,0,cos(a),sin(a),0;0,0,0,-
sin(a),cos(a),0;0,0,0,0,0,1];
```

TransformT.m

```
% Coordinates transformation matrix
function T = TransformT(a)
    T = [cos(a),sin(a),0,0;-sin(a),cos(a),0,0;0,0,cos(a),sin(a);0,0,-
sin(a),cos(a)];
```

TransformTB.m

```
% Coordinates transformation matrix
function T = TransformTB(a)
    T = [cos(a),sin(a),0,0,0;-sin(a),cos(a),0,0,0;0,0,cos(a),sin(a),0;0,0,-
sin(a),cos(a),0;0,0,0,0,1];
```

Post-processing:

plotcrack.m

```
% Postprocessing crack
function  plotcrack(ODB,step,ten_limit,comp_limit)

deformscale = 5;
ID = ODB(step).ID;
ID2 = ODB(step).ID2;
eleb = ODB(step).BeamElement.Nodes;
x = ODB(step).Node.X+ODB(step).Node.U;
y = ODB(step).Node.Y+ODB(step).Node.V;
xi = ODB(step).Node.X;
Lb = ODB(step).BeamElement.Length;
Uo = InitialCondition(ODB(step).Node.U,ODB(step).Node.V,...
```

```
   ODB(step).Node.T,ODB(step).Node.Boundary);
U = Reorder(ID,Uo);
n = ODB(step).BeamElement.n;
hf = ODB(step).BeamElement.hf;
zs = ODB(step).BeamElement.zs;
Ec = ODB(step).Material.Ec;
Esteel = ODB(step).Material.Es;
S_cm = zeros(2,n);
fig1 = figure('Position',[100 100 1000 600]);

for i = 1 :size(eleb,1)
   %transfor local coordinates to global coordinates
   xe = [x(eleb(i,1)),x(eleb(i,2))];
   ye = [y(eleb(i,1)),y(eleb(i,2))];
   xei = [xi(eleb(i,1)),xi(eleb(i,2))];
   plotx = [(xei(1)+xei(2))/2*ones(1,n)-abs(xei(1)-
xei(2))/2*0.577350269;(xei(1)+xei(2))/2*ones(1,n)+abs(xei(1)-
xei(2))/2*0.577350269];
   ploty = [-hf;-hf];
   if (xe(2) - xe(1)) >= 0;
      a = 2*pi+atan((ye(2)-ye(1))/(xe(2)-xe(1)));
   else
      a = pi+atan((ye(2)-ye(1))/(xe(2)-xe(1)));
   end

   % extract displacement vector for each nonlinear beam element
   Ue = TransformB(a)*[U(ID(ID2(eleb(i,1),1)));...
      U(ID(ID2(eleb(i,1),2)));...
      U(ID(ID2(eleb(i,1),3)));...
      U(ID(ID2(eleb(i,2),1)));...
      U(ID(ID2(eleb(i,2),2)));...
      U(ID(ID2(eleb(i,2),3)))];

    Ue2 = [U(ID(ID2(eleb(i,1),1)));...
      U(ID(ID2(eleb(i,1),2)));...
      U(ID(ID2(eleb(i,1),3)));...
      U(ID(ID2(eleb(i,2),1)));...
      U(ID(ID2(eleb(i,2),2)));...
      U(ID(ID2(eleb(i,2),3)))];

   le = Lb(i);
```

```
% compute B matrix for 6 Gauss pts within one element
B = zeros(2,6,2);
Xi = [-0.577350269,0.577350269]; % 2 pts Gauss quadrature
W = [1,1]; % weights

N1 = NmatrixBeam(Xi(1),le);          % shape functions of bending
deformation
Ns1 = NmatrixBeam1(Xi(1),le);          % shape functions of
bending slop
Na1 = Na_matrixBeam(Xi(1));          % shape functions of axial
deformation
N2 = NmatrixBeam(Xi(2),le);          % shape functions of bending
deformation
Ns2 = NmatrixBeam1(Xi(2),le);          % shape functions of
bending slop
Na2 = Na_matrixBeam(Xi(2));          % shape functions of axial
deformation

slop = [Ns1*Ue2, Ns2*Ue2]*2/le*deformscale;
xe2 = [Na1*Ue2, Na2*Ue2]*deformscale;
ye2 = [N1*Ue2, N2*Ue2]*deformscale;

if (xe(2) - xe(1)) >= 0;
  plotx2 = [(xei(1)+xei(2))/2-0.577350269*abs(xei(1)-
xei(2))/2+xe2(1)*ones(1,n)+hf*slop(1)/sqrt(1+slop(1)^2);...
  (xei(1)+xei(2))/2+0.577350269*abs(xei(1)-
xei(2))/2+xe2(2)*ones(1,n)+hf*slop(2)/sqrt(1+slop(2)^2)];
else
  plotx2 = [(xei(1)+xei(2))/2-0.577350269*abs(xei(1)-
xei(2))/2+xe2(1)*ones(1,n)-hf*slop(1)/sqrt(1+slop(1)^2);...
  (xei(1)+xei(2))/2+0.577350269*abs(xei(1)-
xei(2))/2+xe2(2)*ones(1,n)-hf*slop(2)/sqrt(1+slop(2)^2)];
end

if slop(1) == 0 && slop(2) == 0
  ploty2 = [ye2(1)*ones(1,n)-hf;ye2(2)*ones(1,n)-hf];
else
  if (ye(2) - ye(1)) >= 0;
    ploty2 = [ye2(1)*ones(1,n)-hf*1/sqrt(1+slop(1)^2)*sign(slop(1));...
      ye2(2)*ones(1,n)-hf*1/sqrt(1+slop(2)^2)*sign(slop(2))];
  else
```

```
        ploty2 =
[ye2(1)*ones(1,n)+hf*1/sqrt(1+slop(1)^2)*sign(slop(1));...
            ye2(2)*ones(1,n)+hf*1/sqrt(1+slop(2)^2)*sign(slop(2))];
      end
   end

   epsilon_e = zeros(2,2);
   for loop3 = 1:2
      B(:,:,loop3) = 1/le*[-1,0,0,1,0,0;0,6*Xi(loop3)/le,3*Xi(loop3)-1,0,-
6*Xi(loop3)/le,3*Xi(loop3)+1];
      epsilon_e(:,loop3) = B(:,:,loop3)*Ue;
   end

   epsilon_0 = epsilon_e(1,:)'*ones(1,n);  % axial strain on the ref axis
(constant through one element)
   culvature = epsilon_e(2,:);
   epsilon_0s = epsilon_e(1,:)'*ones(1,size(zs,2));

   epsilon_c = epsilon_0+[culvature(1)*hf;culvature(2)*hf];    % column
stands for each fiber section, row stands for each integration pts
   epsilon_s = epsilon_0s+[culvature(1)*zs;culvature(2)*zs];

   E = interp1(Ec(2,:),Ec(1,:),epsilon_c);
   Es = interp1(Esteel(2,:),Esteel(1,:),epsilon_s);
   S_c = epsilon_c.*E;

   for loop10 = 1:2
      for loop11 = 1:n
         if epsilon_c(loop10,loop11) <= -comp_limit
            S_cm(loop10,loop11) = 1;

plot(plotx2(loop10,loop11),ploty2(loop10,loop11),'Marker','+','MarkerSize
',6,'MarkerEdgeColor','r');
            hold on;
         elseif epsilon_c(loop10,loop11) >= ten_limit
            S_cm(loop10,loop11) = 2;

plot(plotx2(loop10,loop11),ploty2(loop10,loop11),'Marker','+','MarkerSize
',6,'MarkerEdgeColor','b');
            hold on;
         else
```

```
plot(plotx2(loop10,loop11),ploty2(loop10,loop11),'Marker','+','MarkerSize
',4,'MarkerEdgeColor',[.5,.5,.5]);
            hold on;
         end
      end
   end
end

set(gca, 'XLim', [min(ODB(step).Node.X)-(max(ODB(step).Node.X)-
min(ODB(step).Node.X))/10,...
   max(ODB(step).Node.X)+(max(ODB(step).Node.X)-
min(ODB(step).Node.X))/10]);
set(gca, 'YLim', [min(ODB(step).Node.Y)-(max(ODB(step).Node.Y)-
min(ODB(step).Node.Y))/.1,...
   max(ODB(step).Node.Y)+(max(ODB(step).Node.Y)-
min(ODB(step).Node.Y))/.1]);
```

plotdisp.m

```
% plot displacement and tendon stress
function  plotdisp(ODB,step)
elet = ODB(step).TrussElement.Nodes;
eleb = ODB(step).BeamElement.Nodes;
eleemb = ODB(step).EmbeddingElement.Nodes;
xi = ODB(1).Node.X;
yi = ODB(1).Node.Y;

ID = ODB(step).ID;
ID2 = ODB(step).ID2;
Lt = ODB(1).TrussElement.Length;
U = ODB(step).Node.U;
V = ODB(step).Node.V;

x2 = ODB(1).Node.X;
y2 = ODB(1).Node.Y;

deformscale = 5;
nplot = 10;
figure(10)

e_truss = zeros(size(ODB(step).TrussElement.Nodes,1),1);
```

```
s_truss = zeros(size(ODB(step).TrussElement.Nodes,1),1);
integ_pt_truss = zeros(size(ODB(step).TrussElement.Nodes,1),1);
leng_truss_initial =
ElementLength(ODB(step).TrussElement.Nodes,ODB(step).Node.X,ODB
(step).Node.Y);
leng_truss_current =
ElementLength(ODB(step).TrussElement.Nodes,ODB(step).Node.X+OD
B(step).Node.U,...
   ODB(step).Node.Y+ODB(step).Node.V);
colora = rand;colorb = rand;colorc = rand;
for i = 1:size(elet,1)
  plot1(1,1) = xi(elet(i,1))+deformscale*ODB(step).Node.U(elet(i,1));
  plot1(1,2) = yi(elet(i,1))+deformscale*ODB(step).Node.V(elet(i,1));
  plot1(2,1) = xi(elet(i,2))+deformscale*ODB(step).Node.U(elet(i,2));
  plot1(2,2) = yi(elet(i,2))+deformscale*ODB(step).Node.V(elet(i,2));
  plot(plot1(:,1),plot1(:,2),'--rs','Marker','o','LineWidth',1,...
          'MarkerEdgeColor','k',...
          'MarkerFaceColor',[colora,colorb,colorc]);

% linear geometry
   xe = [x2(elet(i,1)),x2(elet(i,2))];
   ye = [y2(elet(i,1)),y2(elet(i,2))];
   if (xe(2) - xe(1)) >= 0;
      a = 2*pi+atan((ye(2)-ye(1))/(xe(2)-xe(1)));
   else
      a = pi+atan((ye(2)-ye(1))/(xe(2)-xe(1)));
   end

   % extract displacement vector for each nonlinear beam element
   Ue = TransformT(a)*[U(elet(i,1));...
      V(elet(i,1));...
      U(elet(i,2));...
      V(elet(i,2))];
   e_truss(i) = 1/Lt(i)*[-1,0,1,0]*Ue;

  s_truss(i) = e_truss(i)*ODB(step).TrussElement.E(i);

text(plot1(1,1)/2+plot1(2,1)/2,plot1(1,2)/2+plot1(2,2)/2+((max(ODB(step)
.Node.Y)...
```

```
  -
min(ODB(step).Node.Y))/10),num2str(round(s_truss(i))),'FontSize',6,'Col
or','blue')
  text(plot1(1,1)/2+plot1(2,1)/2,plot1(1,2)/2+plot1(2,2)/2-
((max(ODB(step).Node.Y)...
  -
min(ODB(step).Node.Y))/5),num2str(e_truss(i)),'FontSize',6,'Color','red')
  integ_pt_truss(i) =
(xi(elet(i,1))+ODB(step).Node.U(elet(i,1))+xi(elet(i,2))+ODB(step).Node.
U(elet(i,2)))/2;
  hold on;
end

for e = 1:size(eleemb,1)
   % extract element nodal displacements
   de(:,1) =
[ODB(step).Node.U(ODB(step).EmbeddingElement.Nodes(e,1));...
      ODB(step).Node.V(ODB(step).EmbeddingElement.Nodes(e,1));...
      ODB(step).Node.T(ODB(step).EmbeddingElement.Nodes(e,1));...
      ODB(step).Node.U(ODB(step).EmbeddingElement.Nodes(e,2));...
      ODB(step).Node.V(ODB(step).EmbeddingElement.Nodes(e,2));...
      ODB(step).Node.T(ODB(step).EmbeddingElement.Nodes(e,2))];
   % extract element connectivity information
   IENe      = ODB(step).EmbeddingElement.Nodes(e,:);
   xe        = xi(IENe);             % extract element coordinates
   ye        = yi(IENe);             % extract element coordinates
   J   = (xe(2) - xe(1))/2;                                % Jacobian
   plotgauss= (-1:2/(nplot-1):1);
   J       = sqrt((xe(2) - xe(1))^2+(ye(2) - ye(1))^2)/2;        % compute
Jacobian
   Le = J*2;                                               % compute
element length

   if (xe(2) - xe(1)) >= 0;
      a = 2*pi+atan((ye(2)-ye(1))/(xe(2)-xe(1)));
   else
      a = pi+atan((ye(2)-ye(1))/(xe(2)-xe(1)));
   end

   initialx = xe(1);
   initialy = ye(1);
```

```
  intervalx = (xe(2)-xe(1))/(nplot-1);
  intervaly = (ye(2)-ye(1))/(nplot-1);

  for i = 1:nplot
      s  = plotgauss(i);
      N  = NmatrixBeam(s,Le);                % shape functions of bending
deformation
      Na  = Na_matrixBeam(s);
      B  = SmatrixBeam(s,Le)*1/J^2;      % second derivative of shape
functions
      localx = Na*TransformB(a)*de;
      localy = N*TransformB(a)*de;
      globalx(i) = cos(a)*localx-sin(a)*localy;
      globaly(i) = sin(a)*localx+cos(a)*localy;
      Initialx(i) = initialx;
      Initialy(i) = initialy;
      initialx = initialx +intervalx;
      initialy = initialy +intervaly;
    end

  p = plot(Initialx+globalx*deformscale,Initialy+globaly*deformscale);
  hold on;
  set(p,'Color',[.5,.5,.5],'LineWidth',0.5)
end

for e = 1:size(eleb,1)
  % extract element nodal displacements
  de(:,1) = [ODB(step).Node.U(ODB(step).BeamElement.Nodes(e,1));...
    ODB(step).Node.V(ODB(step).BeamElement.Nodes(e,1));...
    ODB(step).Node.T(ODB(step).BeamElement.Nodes(e,1));...
    ODB(step).Node.U(ODB(step).BeamElement.Nodes(e,2));...
    ODB(step).Node.V(ODB(step).BeamElement.Nodes(e,2));...
    ODB(step).Node.T(ODB(step).BeamElement.Nodes(e,2))];
  % extract element connectivity information
  IENe      = ODB(step).BeamElement.Nodes(e,:);
  xe        = xi(IENe);                                    % extract
element coordinates
  ye        = yi(IENe);                                    % extract
element coordinates
  J  = (xe(2) - xe(1))/2;                                  % Jacobian
  plotgauss= (-1:2/(nplot-1):1);
```

```
  J      = sqrt((xe(2) - xe(1))^2+(ye(2) - ye(1))^2)/2;           % compute
Jacobian
  Le = J*2;                                                        % compute
element length

  if (xe(2) - xe(1)) >= 0;
     a = 2*pi+atan((ye(2)-ye(1))/(xe(2)-xe(1)));
  else
     a = pi+atan((ye(2)-ye(1))/(xe(2)-xe(1)));
  end

  initialx = xe(1);
  initialy = ye(1);
  intervalx = (xe(2)-xe(1))/(nplot-1);
  intervaly = (ye(2)-ye(1))/(nplot-1);

  for i = 1:nplot
       s   = plotgauss(i);
       N   = NmatrixBeam(s,Le);              % shape functions of bending
deformation
       Na  = Na_matrixBeam(s);
       B   = SmatrixBeam(s,Le)*1/J^2;        % second derivative of shape
functions
       localx = Na*TransformB(a)*de;
       localy = N*TransformB(a)*de;
       globalx(i) = cos(a)*localx-sin(a)*localy;
       globaly(i) = sin(a)*localx+cos(a)*localy;
       Initialx(i) = initialx;
       Initialy(i) = initialy;
       initialx = initialx +intervalx;
       initialy = initialy +intervaly;
     end

  p = plot(Initialx+globalx*deformscale,Initialy+globaly*deformscale);
  set(p,'Color',[0,0,0],'LineWidth',1)

%   set proper scale for plots
  set(gca, 'XLim', [min(ODB(step).Node.X)-(max(ODB(step).Node.X)-
min(ODB(step).Node.X))/10,...
     max(ODB(step).Node.X)+(max(ODB(step).Node.X)-
min(ODB(step).Node.X))/10]);
```

```
   set(gca, 'YLim', [min(ODB(step).Node.Y)-(max(ODB(step).Node.Y)-
min(ODB(step).Node.Y))/.2,...
      max(ODB(step).Node.Y)+(max(ODB(step).Node.Y)-
min(ODB(step).Node.Y))/.2]);
end

% label node number
for i = 1:size(ODB(step).Node.X,2)

text(xi(i)+deformscale*ODB(step).Node.U(i)+((max(ODB(step).Node.Y)-
min(ODB(step).Node.Y))/0.5),...

yi(i)+deformscale*ODB(step).Node.V(i),num2str(i),'FontSize',6,'Color','bl
ack')
end

% plot tendon stress strain varations along beam span
figure(20)
subplot(2,1,1)
plot(integ_pt_truss,e_truss)
hold on;
subplot(2,1,2)
plot(integ_pt_truss,s_truss)
hold on;
```

plotstress.m

```
% Post-processing stress contour fo beam
function  plotstress(ODB,step,ten_limit,comp_limit)

deformscale = 5;
ID = ODB(step).ID;
ID2 = ODB(step).ID2;
eleb = ODB(step).BeamElement.Nodes;
x = ODB(step).Node.X+ODB(step).Node.U;
y = ODB(step).Node.Y+ODB(step).Node.V;
xi = ODB(step).Node.X;
Lb = ODB(step).BeamElement.Length;
Uo = InitialCondition(ODB(step).Node.U,ODB(step).Node.V,...
   ODB(step).Node.T,ODB(step).Node.Boundary);
U = Reorder(ID,Uo);
n = ODB(step).BeamElement.n;
hf = ODB(step).BeamElement.hf;
```

```
zs = ODB(step).BeamElement.zs;
Ec = ODB(step).Material.Ec;
Esteel = ODB(step).Material.Es;
S_cm = zeros(2,n);
conx = [];
cony = [];
conz = [];

for i = 1 :size(eleb,1)

   %transfor local coordinates to global coordinates
   xe = [x(eleb(i,1)),x(eleb(i,2))];
   ye = [y(eleb(i,1)),y(eleb(i,2))];
   xei = [xi(eleb(i,1)),xi(eleb(i,2))];
   plotx = [(xei(1)+xei(2))/2*ones(1,n)-abs(xei(1)-
xei(2))/2*0.577350269;(xei(1)+xei(2))/2*ones(1,n)+abs(xei(1)-
xei(2))/2*0.577350269];
   ploty = [-hf;-hf];
   if (xe(2) - xe(1)) >= 0;
      a = 2*pi+atan((ye(2)-ye(1))/(xe(2)-xe(1)));
   else
      a = pi+atan((ye(2)-ye(1))/(xe(2)-xe(1)));
   end

   % extract displacement vector for each nonlinear beam element
   Ue = TransformB(a)*[U(ID(ID2(eleb(i,1),1)));...
      U(ID(ID2(eleb(i,1),2)));...
      U(ID(ID2(eleb(i,1),3)));...
      U(ID(ID2(eleb(i,2),1)));...
      U(ID(ID2(eleb(i,2),2)));...
      U(ID(ID2(eleb(i,2),3)))];

    Ue2 = [U(ID(ID2(eleb(i,1),1)));...
      U(ID(ID2(eleb(i,1),2)));...
      U(ID(ID2(eleb(i,1),3)));...
      U(ID(ID2(eleb(i,2),1)));...
      U(ID(ID2(eleb(i,2),2)));...
      U(ID(ID2(eleb(i,2),3)))];

   le = Lb(i);
```

```
% compute B matrix for 6 Gauss pts within one element
B = zeros(2,6,2);
Xi = [-0.577350269,0.577350269];          % 2 pts Gauss quadrature
W = [1,1];                                % weights

N1 = NmatrixBeam(Xi(1),le);               % shape functions of bending
deformation
Ns1 = NmatrixBeam1(Xi(1),le);             % shape functions of bending
slop
Na1 = Na_matrixBeam(Xi(1));               % shape functions of axial
deformation
N2 = NmatrixBeam(Xi(2),le);               % shape functions of bending
deformation
Ns2 = NmatrixBeam1(Xi(2),le);             % shape functions of bending
slop
Na2 = Na_matrixBeam(Xi(2));               % shape functions of axial
deformation

slop = [Ns1*Ue2, Ns2*Ue2]*2/le*deformscale;
xe2 = [Na1*Ue2, Na2*Ue2]*deformscale;
ye2 = [N1*Ue2, N2*Ue2]*deformscale;

if (xe(2) - xe(1)) >= 0;
   plotx2 = [(xei(1)+xei(2))/2-0.577350269*abs(xei(1)-
xei(2))/2+xe2(1)*ones(1,n)+hf*slop(1)/sqrt(1+slop(1)^2);...
   (xei(1)+xei(2))/2+0.577350269*abs(xei(1)-
xei(2))/2+xe2(2)*ones(1,n)+hf*slop(2)/sqrt(1+slop(2)^2)];
else
   plotx2 = [(xei(1)+xei(2))/2-0.577350269*abs(xei(1)-
xei(2))/2+xe2(1)*ones(1,n)-hf*slop(1)/sqrt(1+slop(1)^2);...
   (xei(1)+xei(2))/2+0.577350269*abs(xei(1)-
xei(2))/2+xe2(2)*ones(1,n)-hf*slop(2)/sqrt(1+slop(2)^2)];
end

if slop(1) == 0 && slop(2) == 0
   ploty2 = [ye2(1)*ones(1,n)-hf;ye2(2)*ones(1,n)-hf];
else
   if (ye(2) - ye(1)) >= 0;
      ploty2 = [ye2(1)*ones(1,n)-hf*1/sqrt(1+slop(1)^2)*sign(slop(1));...
         ye2(2)*ones(1,n)-hf*1/sqrt(1+slop(2)^2)*sign(slop(2))];
   else
```

```
        ploty2 =
[ye2(1)*ones(1,n)+hf*1/sqrt(1+slop(1)^2)*sign(slop(1));...
            ye2(2)*ones(1,n)+hf*1/sqrt(1+slop(2)^2)*sign(slop(2))];
      end
   end

   epsilon_e = zeros(2,2);
   for loop3 = 1:2
      B(:,:,loop3) = 1/le*[-1,0,0,1,0,0;0,6*Xi(loop3)/le,3*Xi(loop3)-1,0,-
6*Xi(loop3)/le,3*Xi(loop3)+1];
      epsilon_e(:,loop3) = B(:,:,loop3)*Ue;
   end
   % axial strain on the ref axis (constant through one element)
   epsilon_0 = epsilon_e(1,:)'*ones(1,n);
   culvature = epsilon_e(2,:);
   epsilon_0s = epsilon_e(1,:)'*ones(1,size(zs,2));
   % column stands for each fiber section, row stands for each integration
pts
   epsilon_c = epsilon_0+[culvature(1)*hf;culvature(2)*hf];
   epsilon_s = epsilon_0s+[culvature(1)*zs;culvature(2)*zs];

   E = interp1(Ec(2,:),Ec(1,:),epsilon_c);
   Es = interp1(Esteel(2,:),Esteel(1,:),epsilon_s);
   S_c = epsilon_c.*E;

   conx = [conx;plotx2];
   cony = [cony;ploty2];
   conz = [conz;S_c];
end

for loop10 = 1:size(eleb,1)*2
   for loop11 = 1:n
      if conz(loop10,loop11) < -comp_limit
         conz(loop10,loop11) = -comp_limit;
      elseif conz(loop10,loop11) > ten_limit
         conz(loop10,loop11) = ten_limit;
      end
   end
end

conx2 = conx';
```

```
cony2 = cony';
conz2 = conz';

figure('Position',[100 100 1000 600]);
contourf(conx2,cony2,conz2,20);
colorbar('east');

set(gca, 'XLim', [min(ODB(step).Node.X)-(max(ODB(step).Node.X)-
min(ODB(step).Node.X))/10,...
   max(ODB(step).Node.X)+(max(ODB(step).Node.X)-
min(ODB(step).Node.X))/10]);
set(gca, 'YLim', [min(ODB(step).Node.Y)-(max(ODB(step).Node.Y)-
min(ODB(step).Node.Y))/.1,...
   max(ODB(step).Node.Y)+(max(ODB(step).Node.Y)-
min(ODB(step).Node.Y))/.1]);
```

Pre-processing and input file (only example 6.7.2 is shown, while other example follows similar formart and input parameters)

Example 6.7.2.m

```
% preprocessor v0.1v generate mesh and input file for PT beam
clear; clc;
format short;
Node = ClassNode;
BeamElement = ClassBeamElement;
TrussElement = ClassTrussElement;
AnchorElement = ClassAnchorElement;
ContactElement = ClassContactElement;
EmbeddingElement = ClassEmbeddingElement;
Material = ClassMaterial;
ODB = ClassODB;

step = 1;
% define material & geometry properties
pt = [5,40;5,0];
% distance between the extreme bottom fiber of corss section and
reference axis
hr = 20;
% number of fiber along vertical direction of the corss section
BeamElement_n = 20;
% area of each steel layer
```

```
BeamElement_As = [0];
% distance of each steel layer to ref axis (minus sign:above ref axis;
postive sign:below ref axis)BeamElement_zs = [15];
[BeamElement_A,BeamElement_S,BeamElement_I,BeamElement_hf,Bea
mElement_Ss,BeamElement_Is]...
    =
Beamsection(pt,BeamElement_n,hr,BeamElement_As,BeamElement_zs);
TrussElement_A = 1.53;
TrussElement_E = 28500000;
EmbeddingElement_A = 450;
EmbeddingElement_E = 4000000;
EmbeddingElement_I = 337500;
AnchorElement_Epsilon = 285000000;
ContactElement_Epsilon = 28500000;
Material.Mu = 0.3;
% unaxial stress-strain   firs coulumn:stress   second column:strain
Material.Concrete = [-705,-0.1;-1128,-0.04;-2692,-0.01;-3600,-0.005;-
4000,-0.003;-3887,-0.002;-2934,-
0.001;0,0;474,0.000162;20,0.0003;10,0.1];
Material.Steel = [-70000,-0.1;-60000,-
0.002069;0,0;60000,0.002069;70000,0.1];
Material.Psteel = [-281147,-0.04;-272273,-0.03;-263321,-0.02;-259418,-
0.016;-252836,-0.012;-243040,-0.01;-232836,-0.009;...
    -216978,-0.008;-195394,-0.007;-169798,-0.006;-142232,-
0.005;0,0;142232,0.005;169798,0.006;195394,0.007;216978,0.008;...
232836,0.009;243040,0.01;252836,0.012;259418,0.016;263321,0.02;2722
73,0.03;281147,0.04];

grid = 0.00011
stress = Material.Concrete(:,1)';
strain = Material.Concrete(:,2)';
Material.Ec =
[interp1(strain,stress,(strain(1):grid:strain(size(strain,2))))./(strain(1):grid:s
train(size(strain,2)));strain(1):grid:strain(size(strain,2))];
Material.Ect =
[diff(interp1(strain,stress,(strain(1):grid:strain(size(strain,2)))))./grid;(strai
n(1)+grid/2):grid:(strain(size(strain,2))-grid/2)];

steel_stress = Material.Steel(:,1)';
steel_strain = Material.Steel(:,2)';
```

```
Material.Es =
[interp1(steel_strain,steel_stress,(steel_strain(1):grid:steel_strain(size(steel
_strain,2)))))./(steel_strain(1):grid:steel_strain(size(steel_strain,2)));steel_st
rain(1):grid:steel_strain(size(steel_strain,2))],
Material.Est =
[diff(interp1(steel_strain,steel_stress,(steel_strain(1):grid:steel_strain(size(
steel_strain,2))))))./grid;(steel_strain(1)+grid/2):grid:(steel_strain(size(steel
_strain,2))-grid/2)];

Psteel_stress = Material.Psteel(:,1)';
Psteel_strain = Material.Psteel(:,2)';
Material.Eps =
[interp1(Psteel_strain,Psteel_stress,(Psteel_strain(1):grid:Psteel_strain(size
(Psteel_strain,2))))./(Psteel_strain(1):grid:Psteel_strain(size(Psteel_strain,2
)));Psteel_strain(1):grid:Psteel_strain(size(Psteel_strain,2))];
Material.Epst =
[diff(interp1(Psteel_strain,Psteel_stress,(Psteel_strain(1):grid:Psteel_strain
(size(Psteel_strain,2))))))./grid;(Psteel_strain(1)+grid/2):grid:(Psteel_strain(
size(Psteel_strain,2))-grid/2)];

% define prestressing method: 1 end jacking; 2 end jacking
% left -- left end jacking; right -- right end jacking; both -- both ends
jacking
jack_method = 'left';
jack_dts1 = 4.371441;              % left jacking pulling out distance
jack_dts2 = 0;                     % right jacking pulling out distance
% define mesh seed
mesh_sd = 20;
% define tendon profile control points and interpolation method
inter_pts = [0,0.1;20,-1.42;40,-2.74;60,-3.96;80,-5.08;100,-6.11;120,-
7.04;140,-7.87;160,-8.6;180,-9.24;...
200,-9.78;220,-10.22;240,-10.56;260,-10.8;280,-10.95;300,-11;320,-
10.95;340,-10.8;360,-10.56;380,-10.22;...
400,-9.78;420,-9.24;440,-8.6;460,-7.87;480,-7.04;500,-6.11;520,-
5.08;540,-3.96;560,-2.74;580,-1.42;600,0.1;];
inter_method = 'cubic';

% define beam geometry
span = 600;

% define EmbbedingElement
```

```
for loop1 = 1:mesh_sd+1
   Node.X(2*loop1-1) =  span/mesh_sd*(loop1-1);
   Node.Y(2*loop1-1) =  0;
   Node.X(2*loop1) = span/mesh_sd*(loop1-1);
   Node.Y(2*loop1) =
interp1(inter_pts(:,1),inter_pts(:,2),span/mesh_sd*(loop1-1),inter_method);
   EmbeddingElement.Nodes(loop1,:) = [2*loop1-1,2*loop1];
   inter_pts2(loop1,1) = span/mesh_sd*(loop1-1);
   inter_pts2(loop1,2) =
interp1(inter_pts(:,1),inter_pts(:,2),span/mesh_sd*(loop1-1),inter_method);
end
EmbeddingElement.Length =
ElementLength(EmbeddingElement.Nodes,Node.X,Node.Y);
EmbeddingElement.A =
EmbeddingElement_A*ones(size(EmbeddingElement.Nodes,1),1);
EmbeddingElement.E =
EmbeddingElement_E*ones(size(EmbeddingElement.Nodes,1),1);
EmbeddingElement.I =
EmbeddingElement_I*ones(size(EmbeddingElement.Nodes,1),1);

% define BeamElement
for loop2 = 1:mesh_sd
   Node.Y(2*loop2-1) = 0;
   BeamElement.Nodes(loop2,:) = [2*loop2-1,2*loop2+1];
end
BeamElement.Length =
ElementLength(BeamElement.Nodes,Node.X,Node.Y);
BeamElement.A = BeamElement_A;
BeamElement.S = BeamElement_S;
BeamElement.I = BeamElement_I;
BeamElement.hf = BeamElement_hf;
BeamElement.n = BeamElement_n;
BeamElement.As = BeamElement_As;
BeamElement.Ss = BeamElement_Ss;
BeamElement.Is = BeamElement_Is;
BeamElement.zs = BeamElement_zs;

% define TrussElement
if strcmp(jack_method,'left') == 1
   anchor_pts = 1;
elseif strcmp(jack_method,'right') == 1
```

```
   anchor_pts = 1;
elseif strcmp(jack_method,'both') == 1
   anchor_pts = 2;
else
   anchor_pts = 0;
end

for loop3 = 1:mesh_sd+1
   if strcmp(jack_method,'left') == 1 && loop3 == 1
      Node.X((mesh_sd+1)*2+1) = jack_dts1;
      Node.Y((mesh_sd+1)*2+1) =
interp1(inter_pts2(:,1),inter_pts2(:,2),jack_dts1);
      TrussElement.Nodes(1,:) = [(mesh_sd+1)*2+1,(mesh_sd+1)*2+2];

   elseif strcmp(jack_method,'left') == 1 && loop3 == mesh_sd+1
      Node.X((mesh_sd+1)*2+loop3) =
(Node.X(BeamElement.Nodes(loop3-
1,1))+Node.X(BeamElement.Nodes(loop3-1,2)))/2;
      Node.Y((mesh_sd+1)*2+loop3) =
interp1(inter_pts2(:,1),inter_pts2(:,2),Node.X((mesh_sd+1)*2+loop3));
      TrussElement.Nodes(loop3,:) =
[(mesh_sd+1)*2+loop3,(mesh_sd+1)*2];

   elseif strcmp(jack_method,'right') == 1 && loop3 == 1
      Node.X((mesh_sd+1)*2+1) =
(Node.X(BeamElement.Nodes(1,1))+Node.X(BeamElement.Nodes(1,2)))/
2;
      Node.Y((mesh_sd+1)*2+1) =
interp1(inter_pts2(:,1),inter_pts2(:,2),Node.X((mesh_sd+1)*2+1));
      TrussElement.Nodes(1,:) = [2,(mesh_sd+1)*2+1];

   elseif strcmp(jack_method,'right') == 1 && loop3 == mesh_sd+1
      Node.X((mesh_sd+1)*2+loop3) = span-jack_dts2;
      Node.Y((mesh_sd+1)*2+loop3) =
interp1(inter_pts2(:,1),inter_pts2(:,2),span-jack_dts2);
      TrussElement.Nodes(loop3,:) = [(mesh_sd+1)*2+loop3-
1,(mesh_sd+1)*2+loop3];

   elseif strcmp(jack_method,'both') == 1 && loop3 == 1
      Node.X((mesh_sd+1)*2+1) = jack_dts1;
```

```
      Node.Y((mesh_sd+1)*2+1) =
interp1(inter_pts2(:,1),inter_pts2(:,2),jack_dts1);
      TrussElement.Nodes(1,:) = [(mesh_sd+1)*2+1,(mesh_sd+1)*2+2];

   else
      if strcmp(jack_method,'left') == 1
         Node.X((mesh_sd+1)*2+loop3) =
(Node.X(BeamElement.Nodes(loop3-
1,1))+Node.X(BeamElement.Nodes(loop3-1,2)))/2;
         TrussElement.Nodes(loop3,:) =
[(mesh_sd+1)*2+loop3,(mesh_sd+1)*2+loop3+1];
      elseif strcmp(jack_method,'right') == 1
         Node.X((mesh_sd+1)*2+loop3) =
(Node.X(BeamElement.Nodes(loop3,1))+Node.X(BeamElement.Nodes(lo
op3,2)))/2;
         TrussElement.Nodes(loop3,:) = [(mesh_sd+1)*2+loop3-
1,(mesh_sd+1)*2+loop3];
      elseif strcmp(jack_method,'both') == 1
         Node.X((mesh_sd+1)*2+loop3) =
(Node.X(BeamElement.Nodes(loop3-
1,1))+Node.X(BeamElement.Nodes(loop3-1,2)))/2;
         TrussElement.Nodes(loop3,:) =
[(mesh_sd+1)*2+loop3,(mesh_sd+1)*2+loop3+1];
      end
      Node.Y((mesh_sd+1)*2+loop3) =
interp1(inter_pts2(:,1),inter_pts2(:,2),Node.X((mesh_sd+1)*2+loop3));
   end
end
if strcmp(jack_method,'both') == 1
   Node.X((mesh_sd+1)*2+mesh_sd+2) = span-jack_dts2;
   Node.Y((mesh_sd+1)*2+mesh_sd+2) =
interp1(inter_pts2(:,1),inter_pts2(:,2),span-jack_dts2);
end
TrussElement.Length =
ElementLength(TrussElement.Nodes,Node.X,Node.Y);
TrussElement.A = TrussElement_A*ones(size(TrussElement.Nodes,1),1);
TrussElement.E = TrussElement_E*ones(size(TrussElement.Nodes,1),1);

% define ContactElement
for loop4 = 1:mesh_sd
   ContactElement.Nodes(loop4,1) = TrussElement.Nodes(loop4,2);
```

```
    ContactElement.Nodes(loop4,2) = EmbeddingElement.Nodes(loop4,2);
    ContactElement.Nodes(loop4,3) =
EmbeddingElement.Nodes(loop4+1,2);
    ContactElement.Nodes(loop4,4) = TrussElement.Nodes(loop4,1);
    ContactElement.Nodes(loop4,5) = TrussElement.Nodes(loop4+1,2);
end
ContactElement.Epsilon = ContactElement_Epsilon;

% define AnchorElement
if strcmp(jack_method,'left') == 1
    AnchorElement.Nodes = [TrussElement.Nodes(1,1),2];
elseif strcmp(jack_method,'right') == 1
    AnchorElement.Nodes =
[TrussElement.Nodes(mesh_sd+1,2),(mesh_sd+1)*2];
elseif strcmp(jack_method,'both') == 1
    AnchorElement.Nodes =
[TrussElement.Nodes(1,1),2;TrussElement.Nodes(mesh_sd+1,2),(mesh_s
d+1)*2];
end
AnchorElement.Epsilon = AnchorElement_Epsilon;

% define residual force & residual displacement
Node.Fx_residual = zeros(size(Node.X));
Node.Fy_residual = zeros(size(Node.X));
Node.M_residual = zeros(size(Node.X));
Node.U = zeros(size(Node.X));
if strcmp(jack_method,'left') == 1
    Node.U((mesh_sd+1)*2+1) = -jack_dts1;
elseif strcmp(jack_method,'right') == 1
    Node.U((mesh_sd+1)*2+mesh_sd+1) = jack_dts2;
elseif strcmp(jack_method,'both') == 1
    Node.U((mesh_sd+1)*2+1) = -jack_dts1;
    Node.U((mesh_sd+1)*2+mesh_sd+2) = jack_dts2;
end
Node.V = zeros(size(Node.X));
if strcmp(jack_method,'left') == 1
    Node.V((mesh_sd+1)*2+1) = Node.Y(2)-Node.Y((mesh_sd+1)*2+1);
elseif strcmp(jack_method,'right') == 1
    Node.V((mesh_sd+1)*2+mesh_sd+1) = Node.Y((mesh_sd+1)*2)-
Node.Y((mesh_sd+1)*2+mesh_sd+1);
elseif strcmp(jack_method,'both') == 1
```

```
   Node.V((mesh_sd+1)*2+1) = Node.Y(2)-Node.Y((mesh_sd+1)*2+1);
   Node.V((mesh_sd+1)*2+mesh_sd+2) = Node.Y((mesh_sd+1)*2)-
Node.Y((mesh_sd+1)*2+mesh_sd+2);
end
Node.T = zeros(size(Node.X));

% define boundary condition per each node
for loop5 = 1:(mesh_sd+1)*2
   Node.Boundary(loop5,:) = [1,1,1];
end
Node.Boundary(1,:) = [2,2,1];
Node.Boundary((mesh_sd+1)*2-1,:) = [1,2,1];
if strcmp(jack_method,'left') == 1
   Node.Boundary((mesh_sd+1)*2+1,:) = [2,2,0];
elseif strcmp(jack_method,'right') == 1
   Node.Boundary((mesh_sd+1)*2+mesh_sd+1,:) = [2,2,0];
elseif strcmp(jack_method,'both') == 1
   Node.Boundary((mesh_sd+1)*2+1,:) = [2,2,0];
   Node.Boundary((mesh_sd+1)*2+mesh_sd+2,:) = [2,2,0];
end

if strcmp(jack_method,'left') == 1
   for loop6 = 1:mesh_sd+1
      Node.Boundary((mesh_sd+1)*2+loop6,:) = [1,1,0];
   end
   Node.Boundary((mesh_sd+1)*2+1,:) = [2,2,0];

elseif strcmp(jack_method,'right') == 1
   for loop6 = 1:mesh_sd+1
      Node.Boundary((mesh_sd+1)*2+loop6,:) = [1,1,0];
   end
   Node.Boundary((mesh_sd+1)*2+mesh_sd+1,:) = [2,2,0];

elseif strcmp(jack_method,'both') == 1
   for loop6 = 1:mesh_sd+2
      Node.Boundary((mesh_sd+1)*2+loop6,:) = [1,1,0];
   end
   Node.Boundary((mesh_sd+1)*2+1,:) = [2,2,0];
   Node.Boundary((mesh_sd+1)*2+mesh_sd+2,:) = [2,2,0] ;
else
   for loop6 = 1:mesh_sd
```

```
      Node.Boundary((mesh_sd+1)*2+loop6,:) = [1,1,0];
   end
end

% define degree of freedom per each node
Node.DOF(1:(mesh_sd+1)*2) = 3*ones((mesh_sd+1)*2,1);
if strcmp(jack_method,'left') == 1 || strcmp(jack_method,'right') == 1
   Node.DOF((mesh_sd+1)*2+1:(mesh_sd+1)*2+mesh_sd+1) =
2*ones(mesh_sd+1,1);
elseif strcmp(jack_method,'both') == 1
   Node.DOF((mesh_sd+1)*2+1:(mesh_sd+1)*2+mesh_sd+2) =
2*ones(mesh_sd+2,1);
else
   Node.DOF((mesh_sd+1)*2+1:(mesh_sd+1)*2+mesh_sd) =
2*ones(mesh_sd,1);
end

ODB(step).Node = Node;
ODB(step).BeamElement = BeamElement;
ODB(step).TrussElement = TrussElement;
ODB(step).ContactElement = ContactElement;
ODB(step).EmbeddingElement = EmbeddingElement;
ODB(step).AnchorElement = AnchorElement;
ODB(step).Material = Material;
% Number of degrees of freedom on the essential boundary
ODB(step).nd = sum(Node.Boundary(:) == 2);
ODB(step).nnd = size(Node.X,2);
ODB(step).neq = sum(Node.DOF);
ODB(step).anchor = 0;

% Generate NodeMapping & Essential boundary vector
[ODB(step).ID,ODB(step).d_E] =
NodeMapping(ODB(step).nnd,ODB(step).nd,ODB(step).Node.DOF,...

ODB(step).Node.Boundary,[ODB(step).Node.U;ODB(step).Node.V;ODB
(step).Node.T]);
% Generate NodeMapping2
ID2 = NodeMapping2(ODB(step).nnd,Node.DOF);
ODB(step).ID2 = ID2;
% Initiate node-slip array
ODB(step).slip = Slip_array(ODB,step);
```

```
% Call MainSolver3 function to calculate new residual force vector and
ODB(step).Node.U((mesh_sd+1)*2+1) =
0.01*ODB(step).Node.U((mesh_sd+1)*2+1);
ODB(step).Node.V((mesh_sd+1)*2+1) =
0.01*ODB(step).Node.V((mesh_sd+1)*2+1);
temp1 = ODB(step).Node.U;
temp2 = ODB(step).Node.V;

[ODB(step).ID,ODB(step).d_E] =
NodeMapping(ODB(step).nnd,ODB(step).nd,ODB(step).Node.DOF,...

ODB(step).Node.Boundary,[ODB(step).Node.U;ODB(step).Node.V;ODB
(step).Node.T]);
ODB(step).Node.U((mesh_sd+1)*2+1) = 0;
ODB(step).Node.V((mesh_sd+1)*2+1) = 0;
ODB = MainSolver3(ODB,step);

loadloop = 1;
for i = 1:99
  [ODB(2).ID,ODB(2).d_E] =
NodeMapping(ODB(2).nnd,ODB(2).nd,ODB(2).Node.DOF,...

ODB(2).Node.Boundary,[temp1;temp2;zeros(1,size(ODB(1).Node.T,2))]);
  ODB = MainSolver3(ODB,2);
  str = ['step=',num2str(step+1),'
disp=',num2str(ODB(loadloop+1).Node.V(1+mesh_sd))];
  msgbox(str,'replace');
  ODB(2) = ODB(3);
end
plotdisp(ODB,2);

loadloop = 1;
for i = 1:10
  [ODB(2).ID,ODB(2).d_E] =
NodeMapping(ODB(2).nnd,ODB(2).nd,ODB(2).Node.DOF,...
    ODB(2).Node.Boundary,[-.046*temp1;-
.046*temp2;zeros(1,size(ODB(1).Node.T,2))]);
  ODB = MainSolver2(ODB,2);
  plotdisp(ODB,3);
  str = ['step=',num2str(step+1),'
disp=',num2str(ODB(loadloop+1).Node.V(1+mesh_sd))];
```

```
    msgbox(str,'replace');
    ODB(2) = ODB(3);
end
ODB(3) = [];

% prestressing settle down
step = 2;
if strcmp(jack_method,'left') == 1
    ODB(step).Node.Boundary((mesh_sd+1)*2+1,:) = [1,1,0];
elseif strcmp(jack_method,'right') == 1
    ODB(step).Node.Boundary((mesh_sd+1)*2+mesh_sd+1,:) = [1,1,0];
elseif strcmp(jack_method,'both') == 1
    ODB(step).Node.Boundary((mesh_sd+1)*2+1,:) = [1,1,0];
    ODB(step).Node.Boundary((mesh_sd+1)*2+mesh_sd+2,:) = [1,1,0];
End
% Number of degrees of freedom on the essential boundary
ODB(step).nd = sum(ODB(step).Node.Boundary(:) == 2);
ODB(step).anchor = 1;
% Generate NodeMapping & Essential boundary vector
[ODB(step).ID,ODB(step).d_E] =
NodeMapping(ODB(step).nnd,ODB(step).nd,ODB(step).Node.DOF,...

ODB(step).Node.Boundary,[ODB(step).Node.U;ODB(step).Node.V;ODB
(step).Node.T]);
% Generate NodeMapping2
ID2 = NodeMapping2(ODB(step).nnd,Node.DOF);
ODB(step).ID2 = ID2;
ODB = MainSolver2(ODB,step);
plotdisp(ODB,3);
```